Student Solutions Manual

for

Strang's

Linear Algebra and Its Applications

Fourth Edition

Gilbert Strang
Massachusetts Institute of Technology

Brett Coonley
Massachusetts Institute of Technology

Andrew Bulman-Fleming

Australia • Brazil • Japan • Korea • Mexico • Singapore • Spain • United Kingdom • United States

Student Solutions Manual for Strang's Linear Algebra and Its Applications, Fourth Edition
Gilbert Strang, Brett Coonley, Andrew Bulman-Fleming

Cover Image: Judith Laurel Harkness

ISBN-13: 978-0-495-01325-9

ISBN-10: 0-495-01325-0

Brooks/Cole
10 Davis Drive
Belmont, CA 94002-3098
USA

Printed in the United States of America
4 5 6 7 8 20 19 18 17 16

Contents

Chapter 1 Matrices and Gaussian Elimination

Problem Set 1.2, page 9

1.2.1 To solve $x + y = 4$ and $2x - 2y = 4$, we subtract the second equation from twice the first to get $4y = 4$ or $y = 1$. Then from the first equation, $x = 3$. So the lines intersect at $(x, y) = (3, 1)$. Then $3\,(\text{column 1}) + 1\,(\text{column 2}) = (4, 4)$.

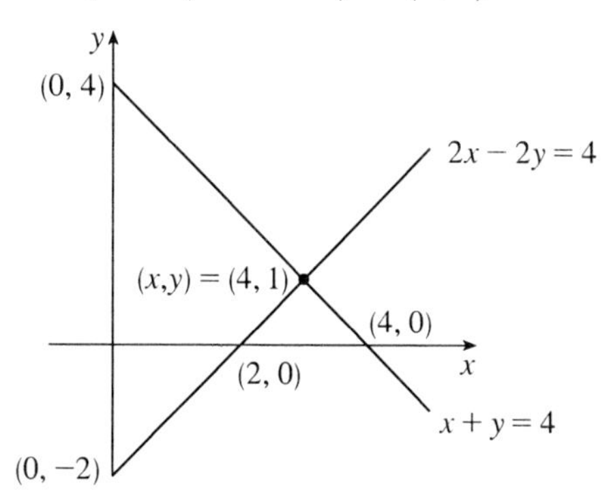

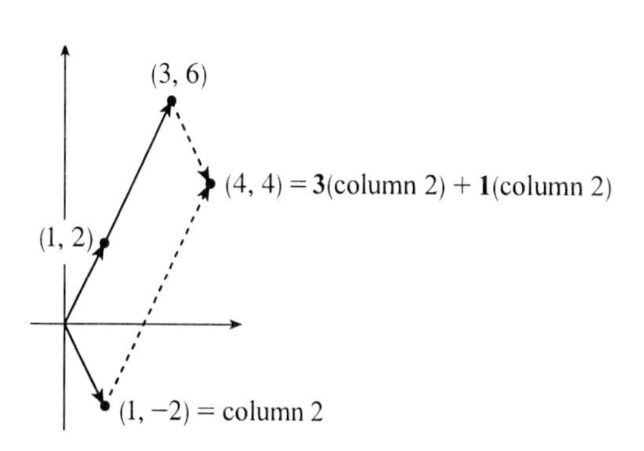

1.2.3 These "planes" intersect in a line in four-dimensional space. In particular, $u + w = 2$ and $u + w + z = 4$ give $z = 2$, and it follows from $u + v + w + z = 6$ that $v = 2$. Thus, the line of intersection is $(c, 2, 2 - c, 2)$ for all real numbers c.
With the plane $u = 1$, the solution is the point $(1, 2, 1, 2)$.
An inconsistent equation such as $u + w = 5$ leaves no solution (no intersection).

1.2.5 $t = 0$, $z = 0$, and $x + y + z + t = 1$ imply that $x + y = 1$. Two such points are $(1, 0, 0, 0)$ and $(0, 1, 0, 0)$.

1.2.7 $u \begin{bmatrix} 1 \\ 2 \\ 3 \end{bmatrix} + v \begin{bmatrix} 1 \\ 0 \\ 1 \end{bmatrix} + w \begin{bmatrix} 1 \\ 3 \\ 4 \end{bmatrix} = \begin{bmatrix} b_1 \\ b_2 \\ b_3 \end{bmatrix} = b$ is equivalent to the three equations

$$\begin{cases} (1)\ u + v + w = b_1 \\ (2)\ 2u + 3w = b_2 \\ (3)\ 3u + v + 4w = b_3 \end{cases}$$

Now (3) − (1) − (2) gives us $b_3 - b_2 - b_1 = 0$, so the plane of intersection is $b = \{(c, d, c + d)$ for all real numbers c and $d\}$. The system is solvable for any b satisfying this condition; for example, $(3, 5, 8)$ or $(1, 2, 3)$. It is not solvable for $b = (3, 5, 7)$ or $b = (1, 2, 2)$.

1.2.9 Column 3 = 2 (column 2) − column 1. If $b = (0, 0, 0)$ then $(u, v, w) = (c, -2c, c)$.

1.2.11 In order to have a line of solutions, the second equation must be a linear multiple of the first. Solving $a \cdot c = 2$, $c \cdot 2 = a$ gives $2c^2 = 2 \quad \Leftrightarrow \quad c = \pm 1 \quad \Leftrightarrow \quad a = \pm 2$. Otherwise, the only solution is the trivial $(0, 0)$.

1.2.13 The row picture has two lines meeting at $(4, 2)$. The column picture has
$4\,(1,1) + 2\,(-2,1) = 4\,(\text{column } 1) + 2\,(\text{column } 2) = (0,6)$.

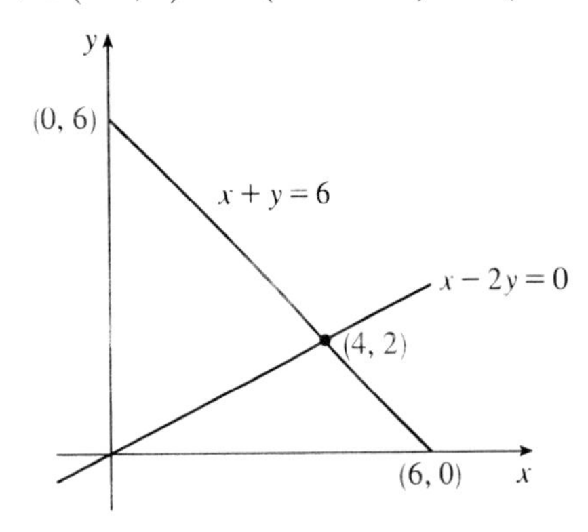

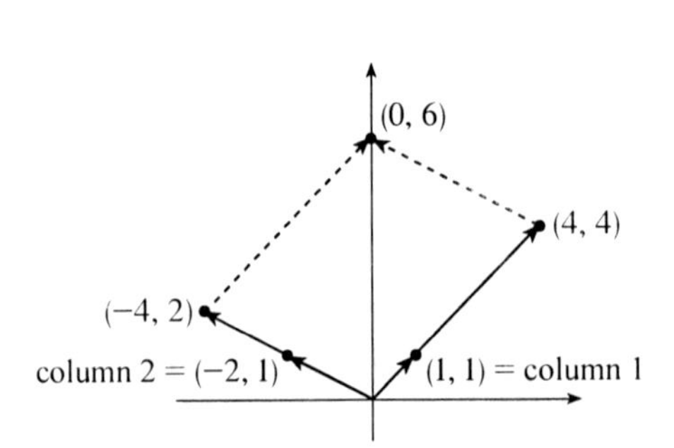

1.2.15 For four linear equations in two unknowns x and y, the row picture shows four *lines*. The column picture is in *four*-dimensional space. The equations have no solution unless the vector on the right-hand side is a combination of *the two columns*.

1.2.17 If x, y, z satisfy the first two equations then they also *satisfy the third equation*. The line $\mathbf{L}$ of solutions contains $\mathbf{v} = (1,1,0)$ and $\mathbf{w} = \left(\frac{1}{2}, 1, \frac{1}{2}\right)$ and
$\mathbf{u} = \frac{1}{2}\mathbf{v} + \frac{1}{2}\mathbf{w} = \left(\frac{3}{4}, 1, \frac{1}{4}\right)$ and all combinations $c\mathbf{v} + d\mathbf{w}$ with $c + d = 1$.

1.2.19 The third column is *equal to the first column*. Two possibilities are $(x, y, z) = (1,1,0)$ and $(0,1,1)$ and you can add any multiple of $(-1,0,1)$; $b = (4,6,c)$ needs $c = 10$ for solvability.

1.2.21 The second plane and row 2 of the matrix and all columns of the matrix are changed. The solution is not changed.

1.2.23 $(u, v, w) = (0,0,1)$, because $1\,(\text{column } 3) = b$.

Problem Set 1.3, page 15

1.3.1 Multiply by $\ell = \frac{10}{2} = 5$ and subtract to find the triangular system $\begin{cases} 2x + 3y = 1 \\ \quad\;\; -6y = 6 \end{cases}$

The pivots are 2 and -6.

1.3.3 Subtract $-\frac{1}{2}$ times equation 1 (or add $\frac{1}{2}$ times equation 1). The new second equation is $3y = 3$. Then $y = 1$ and $x = 5$. If the right side changes sign, so does the solution: $(x, y) = (-5, -1)$.

1.3.5 $6x + 4y$ is 2 times $3x + 2y$. There is no solution unless the right side is $2 \cdot 10 = 20$. Then all points on the line $3x + 2y = 10$ are solutions, including $(0, 5)$ and $(4, -1)$.

1.3.7 If $a = 2$, then elimination must fail. The equations have no solution. If $a = 0$, then elimination stops for a row exchange. Then $3y = -3$ gives $y = -1$ and $4x + 6y = 6$ gives $x = 3$.

1.3.9 $6x - 4y$ is 2 times $3x - 2y$. Therefore we need $b_2 = 2b_1$. If that is the case, there are infinitely many solutions; otherwise, there is no solution.

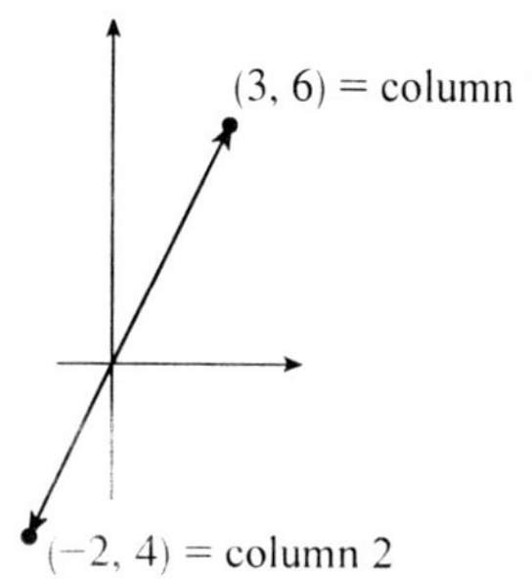

The columns $(3, 6)$ and $(-2, -4)$ are on the same line.

1.3.11 $\begin{cases} 2x - 3y = 3 \\ 4x - 5y + z = 7 \\ 2x - y - 3z = 5 \end{cases} \Rightarrow \begin{cases} \textcircled{2}x - 3y = 3 & \\ \textcircled{1}y + z = 1 & \text{Subtract } 2 \times \text{row 1 from row 2} \\ 2y - 3z = 2 & \text{Subtract } 1 \times \text{row 1 from row 3} \end{cases}$

$\Rightarrow \begin{cases} \textcircled{2}x - 3y = 3 & \\ \textcircled{1}y + z = 1 & \\ \textcircled{-5}z = 0 & \text{Subtract } 2 \times \text{row 2 from row 3} \end{cases}$

Now back-substitution gives $(x, y, z) = (3, 1, 0)$.

1.3.13 After subtracting the first equation from the second, the second pivot position contains $-2 - b$. If $-2 - b = 0 \Leftrightarrow b = -2$ we exchange row 2 and row 3. If $-2 - b = 1 \Leftrightarrow b = -1$, we have the singular case in which the second equation is $-y - z = 0$. In this case, the solutions are of the form $(c, c, -c)$, e.g. $(1, 1, -1)$.

1.3.15 If row 1 = row 2, then row 2 is zero after the first step, so we exchange the zero row with row 3. Thus there is no *third* pivot.

If column 1 = column 2, there is no *second* pivot.

1.3.17 Row 2 becomes $3y - 4z = 5$ and row 3 becomes $(q + 4)z = t - 5$. If $q = -4$, the system is singular (there is no third pivot). Then if $t = 5$, the third equation is $0 = 0$ and there are infinitely many solutions. Choosing $z = 1$, the equation $3y - 4z = 5$ gives $y = 3$ and equation 1 gives $x = -9$, so the solution is $(-9, 3, 1)$.

1.3.19 The system is singular if row 3 is a *linear combination* of the first two rows. From the end view, the three planes form a triangle. This happens if row 1 + row 2 = row 3 on the left-hand side but not the right-hand side; for example, $x + y + z = 0$, $x - 2y - z = 1$, $2x - y = 9$. No two planes are parallel, but there is still no solution.

1.3.21 In Problem 20, the first pivot is 2, the second pivot is $2 - \frac{1}{2} = \frac{3}{2}$, the third pivot is $2 - \frac{2}{3} = \frac{4}{3}$, the fourth pivot is $2 - \frac{3}{4} = \frac{5}{4}$. We see that the fifth pivot is $\frac{6}{5}$, and in general the nth pivot is $\frac{n+1}{n}$.

1.3.23 $\begin{bmatrix} 1 & 1 & 1 & 2 \\ 1 & 3 & 3 & 0 \\ 1 & 3 & 5 & 2 \end{bmatrix} \to \begin{bmatrix} 1 & 1 & 1 & 2 \\ 0 & 2 & 2 & -2 \\ 0 & 2 & 4 & 0 \end{bmatrix} \to \begin{bmatrix} 1 & 1 & 1 & 2 \\ 0 & 2 & 2 & -2 \\ 0 & 0 & 2 & 2 \end{bmatrix}$. The triangular system is

$\begin{cases} u + v + w = 2 \\ 2v + 2w = -2 \\ 2w = 2 \end{cases}$ and the solution is $(3, -2, 1)$.

1.3.25 $\begin{bmatrix} 1 & 1 & 1 & -2 \\ 3 & 3 & -1 & 6 \\ 1 & -1 & 1 & -1 \end{bmatrix} \to \begin{bmatrix} 1 & 1 & 1 & -2 \\ 0 & 0 & -4 & 12 \\ 0 & -2 & 0 & 1 \end{bmatrix}$. At this point we exchange row 2 and row 3:

$\begin{bmatrix} 1 & 1 & 1 & -2 \\ 0 & -2 & 0 & 1 \\ 0 & 0 & -4 & 12 \end{bmatrix}$, which has solution $(u, v, w) = \left(\frac{3}{2}, -\frac{1}{2}, -3\right)$. Changing the coefficient of v in the third equation to 1 would make the system singular, because it would have two equal columns.

1.3.27 $a = 0$ requires a row exchange, but the system is nonsingular: $a = 2$ makes it singular (one pivot, infinite number of solutions); $a = -2$ makes it singular (one pivot, no solution).

1.3.29 The second term $bc + ad$ is $(a + b)(c + d) - ac - bd$, so we need only perform three multiplications: ac, bd, and $(a + b)(c + d)$.

1.3.31 Elimination fails for $a = 2$ (equal columns), for $a = 4$ (equal rows), and for $a = 0$ (zero column).

Problem Set 1.4, page 26

1.4.1 $\begin{bmatrix} 4 & 0 & 1 \\ 0 & 1 & 0 \\ 4 & 0 & 1 \end{bmatrix} \begin{bmatrix} 3 \\ 4 \\ 5 \end{bmatrix} = \begin{bmatrix} 17 \\ 4 \\ 17 \end{bmatrix}$, $\begin{bmatrix} 1 & 0 & 0 \\ 0 & 1 & 0 \\ 0 & 0 & 1 \end{bmatrix} \begin{bmatrix} 5 \\ -2 \\ 3 \end{bmatrix} = \begin{bmatrix} 5 \\ -2 \\ 3 \end{bmatrix}$, and

$\begin{bmatrix} 2 & 0 \\ 1 & 3 \end{bmatrix} \begin{bmatrix} 1 \\ 1 \end{bmatrix} = \begin{bmatrix} 2 \\ 4 \end{bmatrix}$.

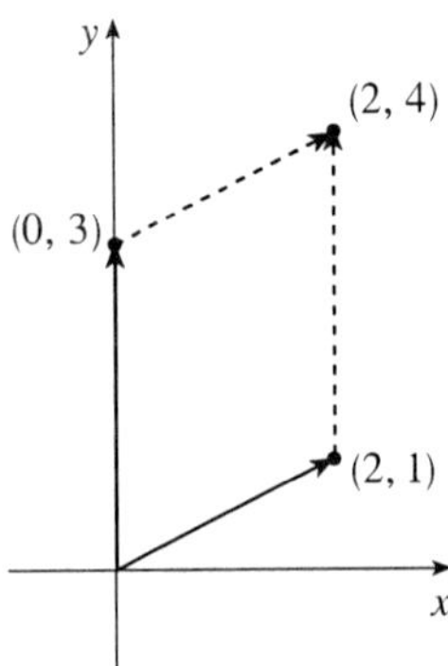

The parallelogram's sides terminate at $(2, 1)$ and $(0, 3)$, so its opposite vertex is $(2, 4)$.

1.4.3 $\begin{bmatrix} 1 & -2 & 7 \end{bmatrix} \begin{bmatrix} 1 \\ -2 \\ 7 \end{bmatrix} = 1^2 + (-2)^2 + 7^2 = 54,$

$\begin{bmatrix} 1 & -2 & 7 \end{bmatrix} \begin{bmatrix} 3 \\ 5 \\ 1 \end{bmatrix} = 1 \cdot 3 + (-2) \cdot 5 + 7 \cdot 1 = 0,$

$\begin{bmatrix} 1 \\ -2 \\ 7 \end{bmatrix} \begin{bmatrix} 3 & 5 & 1 \end{bmatrix} = \begin{bmatrix} 1 \cdot 3 & 1 \cdot 5 & 1 \cdot 1 \\ -2 \cdot 3 & -2 \cdot 5 & -2 \cdot 1 \\ 7 \cdot 3 & 7 \cdot 5 & 7 \cdot 1 \end{bmatrix} = \begin{bmatrix} 3 & 5 & 1 \\ -6 & -10 & -2 \\ 21 & 35 & 7 \end{bmatrix}$.

1.4.5 $Ax = \begin{bmatrix} 3 & -6 & 0 \\ 0 & 2 & -2 \\ 1 & -1 & -1 \end{bmatrix} \begin{bmatrix} 2 \\ 1 \\ 1 \end{bmatrix} = \begin{bmatrix} 0 \\ 0 \\ 0 \end{bmatrix}$, so we know that $x = (2, 1, 1)$ is a solution. Other solutions are $cx = (2c, c, c)$.

1.4.7 (a) $\begin{bmatrix} 1 & 0 & 0 \\ 0 & 2 & 0 \\ 0 & 0 & 7 \end{bmatrix}$ is diagonal (b) $\begin{bmatrix} 1 & 3 & 4 \\ 3 & 2 & 0 \\ 4 & 0 & 7 \end{bmatrix}$ is symmetric

(c) $\begin{bmatrix} 1 & 3 & 4 \\ 0 & 2 & 0 \\ 0 & 0 & 7 \end{bmatrix}$ is triangular (d) $\begin{bmatrix} 0 & 3 & 4 \\ -3 & 0 & 0 \\ -4 & 0 & 0 \end{bmatrix}$ is skew-symmetric

1.4.9 (a) The first pivot is a_{11}. (b) The multiplier is $\ell_{i1} = \dfrac{a_{i1}}{a_{11}}$

(c) The new entry at a_{ij} is $a_{ij} - \dfrac{a_{i1}}{a_{11}} a_{1j}$ (d) The second pivot is $a_{22} - \dfrac{a_{21}}{a_{11}} a_{12}$.

1.4.11 The coefficients of the rows of B in the first row of AB are 2, 1, and 4 (that is, the entries in the first row of A). The first row of AB is thus $\begin{bmatrix} 2b_{11} + b_{21} + 4b_{31} & 2b_{12} + b_{22} + 4b_{32} \end{bmatrix} = \begin{bmatrix} 6 & 3 \end{bmatrix}$.

1.4.13 $A = \begin{bmatrix} 0 & 1 \\ -1 & 0 \end{bmatrix}$; $B = \begin{bmatrix} 0 & 1 \\ 0 & 0 \end{bmatrix}$; $C = \begin{bmatrix} 0 & 1 \\ 1 & 0 \end{bmatrix}$ with $D = A$, giving

$$CD = \begin{bmatrix} 0 & 1 \\ 1 & 0 \end{bmatrix} \begin{bmatrix} 0 & 1 \\ -1 & 0 \end{bmatrix} = \begin{bmatrix} -1 & 0 \\ 0 & 1 \end{bmatrix} \text{ and}$$

$$DC = \begin{bmatrix} 0 & 1 \\ -1 & 0 \end{bmatrix} \begin{bmatrix} 0 & 1 \\ 1 & 0 \end{bmatrix} = \begin{bmatrix} 1 & 0 \\ 0 & -1 \end{bmatrix} = -CD;\ E = F = \begin{bmatrix} 1 & -1 \\ 1 & -1 \end{bmatrix}.$$

1.4.15 $AB_1 = B_1 A \Leftrightarrow \begin{bmatrix} a & 0 \\ c & 0 \end{bmatrix} = \begin{bmatrix} a & b \\ 0 & 0 \end{bmatrix}$, so $b = c = 0$.

$AB_2 = B_2 A \Leftrightarrow \begin{bmatrix} 0 & a \\ c & 0 \end{bmatrix} = \begin{bmatrix} c & d \\ 0 & 0 \end{bmatrix}$, so $a = d$. Thus, $A = \begin{bmatrix} a & 0 \\ 0 & a \end{bmatrix} = aI$.

1.4.17 $A(A+B) + B(A+B)$, $(A+B)(B+A)$, and $A^2 + AB + BA + B^2$ always equal $(A+B)^2$. However, $A^2 + 2AB + B^2 = (A+B)^2$ only if A and B happen to commute.

1.4.19 $\begin{bmatrix} a & b \\ c & d \end{bmatrix} \begin{bmatrix} p & q \\ r & s \end{bmatrix} = \begin{bmatrix} a \\ c \end{bmatrix} \begin{bmatrix} p & q \end{bmatrix} + \begin{bmatrix} b \\ d \end{bmatrix} \begin{bmatrix} r & s \end{bmatrix} = \begin{bmatrix} ap + br & aq + bs \\ cp + dr & cq + ds \end{bmatrix}$.

1.4.21 $A^2 = \begin{bmatrix} \frac{1}{2} & \frac{1}{2} \\ \frac{1}{2} & \frac{1}{2} \end{bmatrix} \begin{bmatrix} \frac{1}{2} & \frac{1}{2} \\ \frac{1}{2} & \frac{1}{2} \end{bmatrix} = \begin{bmatrix} \frac{1}{2} & \frac{1}{2} \\ \frac{1}{2} & \frac{1}{2} \end{bmatrix}$. It follows that $A^k = A$.

$$B^2 = \begin{bmatrix} 1 & 0 \\ 0 & -1 \end{bmatrix} \begin{bmatrix} 1 & 0 \\ 0 & -1 \end{bmatrix} = \begin{bmatrix} 1 & 0 \\ 0 & 1 \end{bmatrix},$$

$$B^3 = \begin{bmatrix} 1 & 0 \\ 0 & 1 \end{bmatrix} \begin{bmatrix} 1 & 0 \\ 0 & -1 \end{bmatrix} = \begin{bmatrix} 1 & 0 \\ 0 & -1 \end{bmatrix} = B. \text{ In general, } B^k = \begin{bmatrix} 1 & 0 \\ 0 & (-1)^k \end{bmatrix}.$$

$$C^2 = \begin{bmatrix} \frac{1}{2} & -\frac{1}{2} \\ \frac{1}{2} & -\frac{1}{2} \end{bmatrix} \begin{bmatrix} \frac{1}{2} & -\frac{1}{2} \\ \frac{1}{2} & -\frac{1}{2} \end{bmatrix} = \begin{bmatrix} 0 & 0 \\ 0 & 0 \end{bmatrix}, \text{ so } C^k = \begin{bmatrix} 0 & 0 \\ 0 & 0 \end{bmatrix},\ k \neq 1.$$

1.4.23 $E_{32}E_{21}b = (1, -5, -35)$, but $E_{21}E_{32}b = (1, -5, \mathbf{0})$. So row 3 feels no effect from row 1.

1.4.25 Changing a_{33} from 7 to 11 will change the third pivot from 5 to 9. Changing a_{33} from 7 to 2 will change the pivot from 5 to 0 (no pivot).

1.4.27 To reverse E_{31}, *add* 7 times row 1 to row 3. The matrix is $R_{31} = \begin{bmatrix} 1 & 0 & 0 \\ 0 & 1 & 0 \\ 7 & 0 & 1 \end{bmatrix}$.

$$E_{31}R_{31} = \begin{bmatrix} 1 & 0 & 0 \\ 0 & 1 & 0 \\ -7 & 0 & 1 \end{bmatrix}\begin{bmatrix} 1 & 0 & 0 \\ 0 & 1 & 0 \\ 7 & 0 & 1 \end{bmatrix} = \begin{bmatrix} 1 & 0 & 0 \\ 0 & 1 & 0 \\ 0 & 0 & 1 \end{bmatrix}, \text{ as expected.}$$

1.4.29 (a) $E_{13} = \begin{bmatrix} 1 & 0 & 1 \\ 0 & 1 & 0 \\ 0 & 0 & 1 \end{bmatrix}$ adds row 3 to row 1.

(b) $\begin{bmatrix} 1 & 0 & 1 \\ 0 & 1 & 0 \\ 1 & 0 & 1 \end{bmatrix}$ adds row 1 to row 3 and simultaneously adds row 3 to row 1.

(c) $E_{31}E_{13} = \begin{bmatrix} 2 & 0 & 1 \\ 0 & 1 & 0 \\ 1 & 0 & 1 \end{bmatrix}$ adds row 1 to row 3, then adds the resulting row 3 to row 1. Testing on

the identity matrix: $\begin{bmatrix} 1 & 0 & 0 \\ 0 & 1 & 0 \\ 0 & 0 & 1 \end{bmatrix} \to \begin{bmatrix} 1 & 0 & 0 \\ 0 & 1 & 0 \\ 1 & 0 & 1 \end{bmatrix} \to \begin{bmatrix} 2 & 0 & 1 \\ 0 & 1 & 0 \\ 1 & 0 & 1 \end{bmatrix}$.

1.4.31 E_{21} has $\ell_{21} = \frac{1}{2}$, E_{32} has $\ell_{32} = -\frac{2}{3}$, E_{43} has $\ell_{43} = -\frac{3}{4}$. Otherwise, the E's match I.

1.4.33 The condition that the parabola passes through the point (x, y) is equivalent to the equation

$\begin{bmatrix} 1 & x & x^2 \end{bmatrix}\begin{bmatrix} a \\ b \\ c \end{bmatrix} = y$. We can express the condition that it passes through all of the given

points by the matrix equation $\begin{bmatrix} 1 & 1 & 1 \\ 1 & 2 & 4 \\ 1 & 3 & 9 \end{bmatrix}\begin{bmatrix} a \\ b \\ c \end{bmatrix} = \begin{bmatrix} 4 \\ 8 \\ 14 \end{bmatrix}$. Using elimination, we find

$\begin{bmatrix} 1 & 1 & 1 & 4 \\ 1 & 2 & 4 & 8 \\ 1 & 3 & 9 & 14 \end{bmatrix} \to \begin{bmatrix} 1 & 1 & 1 & 4 \\ 0 & 1 & 3 & 4 \\ 0 & 0 & 2 & 2 \end{bmatrix}$; back-substitution gives $(a, b, c) = (2, 1, 1)$.

1.4.35 (a) Each column of EB is E times a column of B.

(b) $EB = \begin{bmatrix} 1 & 0 \\ 1 & 1 \end{bmatrix}\begin{bmatrix} 1 & 2 & 4 \\ 1 & 2 & 4 \end{bmatrix} = \begin{bmatrix} 1 & 2 & 4 \\ 2 & 4 & 8 \end{bmatrix}$. Rows of EB are combinations of rows of B, so they are multiples of $\begin{bmatrix} 1 & 2 & 4 \end{bmatrix}$.

1.4.37 The third component of Ax is $(\text{row } 3) \cdot x$ or $\sum a_{3j}x_j$. $(A^2)_{11} = (\text{row } 1)\,(\text{column } 1) = \sum a_{1j}a_{j1}$.

1.4.39 $BA = 3I$ is 5 by 5, $AB = 5I$ is 3 by 3, $ABD = 5D$ is 3 by 1. DBA and $A(B + C)$ are undefined.

1.4.41 (a) $B = 4I$ (b) $B = 0$ (c) $B = \begin{bmatrix} 0 & 0 & 1 \\ 0 & 1 & 0 \\ 1 & 0 & 0 \end{bmatrix}$ (d) Every row of B is $1, 0, 0$.

1.4.43 (a) mn (every entry) (b) mnp (c) n^3 (this is n^2 dot products)

1.4.45 $\begin{bmatrix} 1 \\ 2 \\ 2 \end{bmatrix}\begin{bmatrix} 3 & 3 & 0 \end{bmatrix} + \begin{bmatrix} 0 \\ 4 \\ 1 \end{bmatrix}\begin{bmatrix} 1 & 2 & 1 \end{bmatrix} = \begin{bmatrix} 3 & 3 & 0 \\ 6 & 6 & 0 \\ 6 & 6 & 0 \end{bmatrix} + \begin{bmatrix} 0 & 0 & 0 \\ 4 & 8 & 4 \\ 1 & 2 & 1 \end{bmatrix} = \begin{bmatrix} 3 & 3 & 0 \\ 10 & 14 & 4 \\ 7 & 8 & 1 \end{bmatrix}$.

1.4.47 A times B is $A\left[\,|\;|\;|\;|\,\right]$, $\left[\begin{smallmatrix}\text{———}\\\text{———}\\\text{———}\end{smallmatrix}\right]B$, $\left[\begin{smallmatrix}\text{———}\\\text{———}\\\text{———}\end{smallmatrix}\right]\left[\,|\;|\;|\;|\,\right]$, $\left[\,|\;|\;|\;|\,\right]\left[\begin{smallmatrix}\text{———}\\\text{———}\\\text{———}\end{smallmatrix}\right]$.

1.4.49 The $(2,2)$ block $S = D - CA^{-1}B$ is the Schur complement.

1.4.51 If $X = \begin{bmatrix} x_1 & x_2 & x_3 \end{bmatrix}$, then $AX = A\begin{bmatrix} x_1 & x_2 & x_3 \end{bmatrix} = \begin{bmatrix} Ax_1 & Ax_2 & Ax_3 \end{bmatrix} = \begin{bmatrix} 1 & 0 & 0 \\ 0 & 1 & 0 \\ 0 & 0 & 1 \end{bmatrix} = I$.

1.4.53 $A\begin{bmatrix} 1 & 1 \\ 1 & 1 \end{bmatrix} = \begin{bmatrix} a+b & a+b \\ c+d & c+d \end{bmatrix}$ and $\begin{bmatrix} 1 & 1 \\ 1 & 1 \end{bmatrix}A = \begin{bmatrix} a+c & b+d \\ a+c & b+d \end{bmatrix}$, so A commutes with $\begin{bmatrix} 1 & 1 \\ 1 & 1 \end{bmatrix}$ when $b = c$ and $a = d$; that is, when A is of the form $\begin{bmatrix} a & b \\ b & a \end{bmatrix}$.

1.4.55 $2x + 3y + z + 5t = 8$ is $Ax = b$ with the 1 by 4 matrix $A = \begin{bmatrix} 2 & 3 & 1 & 5 \end{bmatrix}$.

1.4.57 The dot product $\begin{bmatrix} 1 & 4 & 5 \end{bmatrix}\begin{bmatrix} x \\ y \\ z \end{bmatrix} = (1 \text{ by } 3)(3 \text{ by } 1)$ is zero for points (x, y, z) on a *plane* $x + 4y + 5z = 0$ perpendicular to the vector $(1, 4, 5)$ in three dimensions. The columns of A are *one*-dimensional vectors.

1.4.59 $A * v = \begin{bmatrix} 3 & 4 & 5 \end{bmatrix}'$ and $v' * v = 3^2 + 4^2 + 5^2 = 50$; $v * A$ gives an error message.

1.4.61 $M = \begin{bmatrix} 8 & 3 & 4 \\ 1 & 5 & 9 \\ 6 & 7 & 2 \end{bmatrix} = \begin{bmatrix} 5+u & 5-u+v & 5-v \\ 5-u-v & 5 & 5+u+v \\ 5+v & 5+u-v & 5-u \end{bmatrix}$. $M_3(1,1,1) = (15, 15, 15)$ and

$$\begin{bmatrix} 1 & 1 & 1 \end{bmatrix}\begin{bmatrix} 8 & 3 & 4 \\ 1 & 5 & 9 \\ 6 & 7 & 2 \end{bmatrix} = \begin{bmatrix} 15 & 15 & 15 \end{bmatrix}.$$

Similarly, for any 4 by 4 magic matrix M_4, $M_4(1,1,1,1) = (34, 34, 34, 34)$, because $\sum_{n=1}^{16} n = 136 = 4(34)$.

Problem Set 1.5, page 39

1.5.1 U is nonsingular when no entry on the main diagonal is zero.

1.5.3 $\begin{bmatrix} 1 & 0 & 0 \\ 2 & 1 & 0 \\ -1 & -1 & 1 \end{bmatrix}\begin{bmatrix} 1 & 0 & 0 \\ -2 & 1 & 0 \\ -1 & 1 & 1 \end{bmatrix} = \begin{bmatrix} 1 & 0 & 0 \\ 0 & 1 & 0 \\ 0 & 0 & 1 \end{bmatrix}$; $\begin{bmatrix} 1 & 0 & 0 \\ -2 & 1 & 0 \\ -1 & 1 & 1 \end{bmatrix}\begin{bmatrix} 1 & 0 & 0 \\ 2 & 1 & 0 \\ -1 & -1 & 1 \end{bmatrix} = I$ also.

$(E^{-1}F^{-1}G^{-1})(GFE) = E^{-1}F^{-1}FE = E^{-1}E = I$; also $(GFE)(E^{-1}F^{-1}G^{-1}) = I$.

1.5.5 $LU = \begin{bmatrix} 1 & 0 & 0 \\ 0 & 1 & 0 \\ 3 & 0 & 1 \end{bmatrix}\begin{bmatrix} 2 & 3 & 3 \\ 0 & 5 & 7 \\ 0 & 0 & -1 \end{bmatrix}$; after elimination $\begin{bmatrix} 2 & 3 & 3 \\ 0 & 5 & 7 \\ 0 & 0 & -1 \end{bmatrix}\begin{bmatrix} u \\ v \\ w \end{bmatrix} = \begin{bmatrix} 2 \\ 2 \\ -1 \end{bmatrix}$.

1.5.7 $FGH = \begin{bmatrix} 1 & 0 & 0 & 0 \\ 2 & 1 & 0 & 0 \\ 0 & 2 & 1 & 0 \\ 0 & 0 & 2 & 1 \end{bmatrix}$; $HGF = \begin{bmatrix} 1 & 0 & 0 & 0 \\ 2 & 1 & 0 & 0 \\ 4 & 2 & 1 & 0 \\ 8 & 4 & 2 & 1 \end{bmatrix}$.

1.5.9 (a) A is nonsingular when $d_1 d_2 d_3 \neq 0$.

(b) Suppose $d_3 \neq 0$. We solve $Lc = b$ going downwards: $Lc = b$ gives $c = \begin{bmatrix} 0 \\ 0 \\ 1 \end{bmatrix}$. Then

$$\begin{bmatrix} d_1 & -d_1 & 0 \\ 0 & d_2 & -d_2 \\ 0 & 0 & d_3 \end{bmatrix} \begin{bmatrix} u \\ v \\ w \end{bmatrix} = \begin{bmatrix} 0 \\ 0 \\ 1 \end{bmatrix} \text{ gives } x = \begin{bmatrix} 1/d_3 \\ 1/d_3 \\ 1/d_3 \end{bmatrix}.$$

1.5.11 $Lc = b$ going downwards gives $c = \begin{bmatrix} 2 \\ -2 \\ 0 \end{bmatrix}$; $Ux = c$ upwards gives $\begin{bmatrix} u \\ v \\ w \end{bmatrix} = \begin{bmatrix} 5 \\ -2 \\ 0 \end{bmatrix}$.

1.5.13 $A = \begin{bmatrix} 1 & 4 & 2 \\ -2 & -8 & 3 \\ 0 & 1 & 1 \end{bmatrix} \rightarrow \begin{bmatrix} 1 & 4 & 2 \\ 0 & 0 & 7 \\ 0 & 1 & 1 \end{bmatrix}$, so we permute rows 2 and 3: $U = \begin{bmatrix} 1 & 4 & 2 \\ 0 & 1 & 1 \\ 0 & 0 & 7 \end{bmatrix}$ and

$P_{23}A = LU$, where $L = \begin{bmatrix} 1 & 0 & 0 \\ 0 & 1 & 0 \\ -2 & 0 & 1 \end{bmatrix}$. Now $Lc = P_{32}b$ gives $c = \begin{bmatrix} -2 \\ 1 \\ 28 \end{bmatrix}$ and $Ux = c$ gives

$x = \begin{bmatrix} 2 \\ -3 \\ 4 \end{bmatrix}$.

$A = \begin{bmatrix} 0 & 1 & 1 \\ 1 & 1 & 0 \\ 1 & 1 & 1 \end{bmatrix}$, so we permute rows 1 and 2: $\begin{bmatrix} 1 & 1 & 0 \\ 0 & 1 & 1 \\ 1 & 1 & 1 \end{bmatrix} \rightarrow \begin{bmatrix} 1 & 1 & 0 \\ 0 & 1 & 1 \\ 0 & 0 & 1 \end{bmatrix}$ and $P_{12}A = LU$ with

$L = \begin{bmatrix} 1 & 0 & 0 \\ 0 & 1 & 0 \\ 1 & 0 & 1 \end{bmatrix}$. Now $Lc = P_{12}b$ gives $c = \begin{bmatrix} 0 \\ 0 \\ 1 \end{bmatrix}$ and $Ux = c$ gives $x = \begin{bmatrix} 1 \\ -1 \\ 1 \end{bmatrix}$.

1.5.15 $A = \begin{bmatrix} 0 & 1 & 1 \\ 1 & 0 & 1 \\ 2 & 3 & 4 \end{bmatrix} \rightarrow \begin{bmatrix} 1 & 0 & 1 \\ 0 & 1 & 1 \\ 2 & 3 & 4 \end{bmatrix} \rightarrow \begin{bmatrix} 1 & 0 & 1 \\ 0 & 1 & 1 \\ 0 & 3 & 2 \end{bmatrix} \rightarrow \begin{bmatrix} 1 & 0 & 1 \\ 0 & 1 & 1 \\ 0 & 0 & -1 \end{bmatrix}$, so

$PA = \begin{bmatrix} 0 & 1 & 0 \\ 1 & 0 & 0 \\ 0 & 0 & 1 \end{bmatrix} \begin{bmatrix} 0 & 1 & 1 \\ 1 & 0 & 1 \\ 2 & 3 & 4 \end{bmatrix} = \begin{bmatrix} 1 & 0 & 1 \\ 0 & 1 & 1 \\ 2 & 3 & 4 \end{bmatrix}$ and

$LDU = \begin{bmatrix} 1 & 0 & 0 \\ 0 & 1 & 0 \\ 2 & 3 & 1 \end{bmatrix} \begin{bmatrix} 1 & 0 & 0 \\ 0 & 1 & 0 \\ 0 & 0 & -1 \end{bmatrix} \begin{bmatrix} 1 & 0 & 1 \\ 0 & 1 & 1 \\ 0 & 0 & 1 \end{bmatrix} = \begin{bmatrix} 1 & 0 & 1 \\ 0 & 1 & 1 \\ 2 & 3 & 4 \end{bmatrix}$.

If $A = \begin{bmatrix} 1 & 2 & 1 \\ 2 & 4 & 2 \\ 1 & 1 & 1 \end{bmatrix}$, we have $PA = \begin{bmatrix} 1 & 0 & 0 \\ 0 & 0 & 1 \\ 0 & 1 & 0 \end{bmatrix} \begin{bmatrix} 1 & 2 & 1 \\ 2 & 4 & 2 \\ 1 & 1 & 1 \end{bmatrix} = \begin{bmatrix} 1 & 2 & 1 \\ 1 & 1 & 1 \\ 2 & 4 & 2 \end{bmatrix}$ and

$LDU = \begin{bmatrix} 1 & 0 & 0 \\ 1 & 1 & 0 \\ 2 & 0 & 1 \end{bmatrix} \begin{bmatrix} 1 & 0 & 0 \\ 0 & -1 & 0 \\ 0 & 0 & 0 \end{bmatrix} \begin{bmatrix} 1 & 2 & 1 \\ 0 & 1 & 0 \\ 0 & 0 & 0 \end{bmatrix} = \begin{bmatrix} 1 & 2 & 1 \\ 1 & 1 & 1 \\ 2 & 4 & 2 \end{bmatrix}$.

1.5.17 L becomes $\begin{bmatrix} 1 & 0 & 0 \\ 1 & 1 & 0 \\ 2 & 0 & 1 \end{bmatrix}$. MATLAB and other codes use $PA = LU$.

1.5.19 In the first matrix, $a = 4$ leads to a row exchange.

$$A = \begin{bmatrix} 1 & 2 & 0 \\ a & 8 & 3 \\ 0 & b & 5 \end{bmatrix} \to \begin{bmatrix} 1 & 2 & 0 \\ 0 & 8-2a & 3 \\ 0 & b & 5 \end{bmatrix} \to \begin{bmatrix} 1 & 2 & 0 \\ 0 & 8-2a & 3 \\ 0 & 0 & 5-3\left(\frac{b}{8-2a}\right) \end{bmatrix}$$, so the matrix is singular if

$5 - 3\left(\frac{b}{8-2a}\right) = 0 \Leftrightarrow 10a + 3b = 40$.

In the second matrix, $c = 0$ leads to a row exchange and $c = 3$ makes the matrix singular.

1.5.21 $\ell_{31} = 1$, $\ell_{32} = 2$, and $\ell_{33} = 1$. We reverse the steps to recover $x + 3y + 6z = 11$ from $Ux = c$: $(1 \times (x + y + z = 5)) + (2 \times (y + 2z = 2)) + (1 \times (z = 2))$ gives $x + 3y + 6z = 11$.

1.5.23 $$\begin{bmatrix} 1 & & \\ 0 & 1 & \\ 0 & -2 & 1 \end{bmatrix} \begin{bmatrix} 1 & & \\ -2 & 1 & \\ 0 & 0 & 1 \end{bmatrix} A = \begin{bmatrix} 1 & 1 & 1 \\ 0 & 2 & 3 \\ 0 & 0 & -6 \end{bmatrix} = U. \quad A = \begin{bmatrix} 1 & 0 & 0 \\ 2 & 1 & 0 \\ 0 & 2 & 1 \end{bmatrix} U = E_{21}^{-1} E_{32}^{-1} U = LU.$$

1.5.25 In the first matrix, $A_{11} = 1 \cdot d + 0 \cdot 0$, but $d = 0$ is impossible (it is a pivot).

In the second matrix, $\begin{bmatrix} 1 & 1 & 0 \\ 1 & 1 & 2 \\ 1 & 2 & 1 \end{bmatrix} = \begin{bmatrix} 1 & & \\ \ell & 1 & \\ m & n & 1 \end{bmatrix} \begin{bmatrix} d & e & g \\ & f & h \\ & & i \end{bmatrix}$. From the first row, $d = 1$ and $e = 1$.

Thus from the second row, $\ell \cdot d = 1 \Leftrightarrow \ell = 1$, but then $e \cdot \ell + f \cdot 1 = 1 \Leftrightarrow f = 0$, which is impossible.

1.5.27 $A = \begin{bmatrix} 2 & 4 & 8 \\ 0 & 3 & 9 \\ 0 & 0 & 7 \end{bmatrix}$ has $L = I$ and $D = \begin{bmatrix} 2 & & \\ & 3 & \\ & & 7 \end{bmatrix}$; $A = LU$ has $U = A$ (pivots on the diagonal);

$A = LDU$ has $U = D^{-1}A = \begin{bmatrix} 1 & 2 & 4 \\ 0 & 1 & 3 \\ 0 & 0 & 1 \end{bmatrix}$ with 1's on the diagonal.

1.5.29 $$\begin{bmatrix} a & a & a & a \\ a & b & b & b \\ a & b & c & c \\ a & b & c & d \end{bmatrix} = \begin{bmatrix} 1 & & & \\ 1 & 1 & & \\ 1 & 1 & 1 & \\ 1 & 1 & 1 & 1 \end{bmatrix} \begin{bmatrix} a & a & a & a \\ & b-a & b-a & b-a \\ & & c-b & c-b \\ & & & d-c \end{bmatrix}$$. In order to have four pivots, we need the diagonal entries of U to be nonzero; that is, $a \neq 0$, $a \neq b$, $b \neq c$, $c \neq d$.

1.5.31 $$\begin{bmatrix} 1 & 1 & 0 \\ 1 & 2 & 1 \\ 0 & 1 & 2 \end{bmatrix} = \begin{bmatrix} 1 & & \\ 1 & 1 & \\ 0 & 1 & 1 \end{bmatrix} \begin{bmatrix} 1 & 1 & 0 \\ & 1 & 1 \\ & & 1 \end{bmatrix} = LIU.$$ In the general case, L and U stay the same:

$$\begin{bmatrix} a & a & 0 \\ a & a+b & b \\ 0 & b & b+c \end{bmatrix} = \begin{bmatrix} 1 & & \\ 1 & 1 & \\ 0 & 1 & 1 \end{bmatrix} \begin{bmatrix} a & & \\ & b & \\ & & c \end{bmatrix} \begin{bmatrix} 1 & 1 & 0 \\ & 1 & 1 \\ & & 1 \end{bmatrix}.$$

1.5.33 $$\begin{bmatrix} 1 & 0 & 0 \\ 1 & 1 & 0 \\ 1 & 1 & 1 \end{bmatrix} c = \begin{bmatrix} 4 \\ 5 \\ 6 \end{bmatrix} \text{ gives } c = \begin{bmatrix} 4 \\ 1 \\ 1 \end{bmatrix}. \quad \begin{bmatrix} 1 & 1 & 1 \\ 0 & 1 & 1 \\ 0 & 0 & 1 \end{bmatrix} x = \begin{bmatrix} 4 \\ 1 \\ 1 \end{bmatrix} \text{ gives } x = \begin{bmatrix} 3 \\ 0 \\ 1 \end{bmatrix}.$$

$$A = LU = \begin{bmatrix} 1 & 0 & 0 \\ 1 & 1 & 0 \\ 1 & 1 & 1 \end{bmatrix} \begin{bmatrix} 1 & 1 & 1 \\ 0 & 1 & 1 \\ 0 & 0 & 1 \end{bmatrix} = \begin{bmatrix} 1 & 1 & 1 \\ 1 & 2 & 2 \\ 1 & 2 & 3 \end{bmatrix}.$$

1.5.35 The 2 by 2 upper submatrix B has the first two pivots: 2 and 7, because elimination on A starts in the upper left corner with elimination on B.

1.5.37 $\begin{bmatrix} 1 & 1 & 1 & 1 & 1 \\ 1 & 2 & 3 & 4 & 5 \\ 1 & 3 & 6 & 10 & 15 \\ 1 & 4 & 10 & 20 & 35 \\ 1 & 5 & 15 & 35 & 70 \end{bmatrix} = \begin{bmatrix} 1 & & & & \\ 1 & 1 & & & \\ 1 & 2 & 1 & & \\ 1 & 3 & 3 & 1 & \\ 1 & 4 & 6 & 4 & 1 \end{bmatrix} \begin{bmatrix} 1 & 1 & 1 & 1 & 1 \\ & 1 & 2 & 3 & 4 \\ & & 1 & 3 & 6 \\ & & & 1 & 4 \\ & & & & 1 \end{bmatrix}.$

Pascal's triangle appears in both L and U. MATLAB's lu code will wreck the pattern. chol does no row exchanges for symmetric matrices with positive pivots.

1.5.39 Each new *right side* costs only n^2 steps, compared to $\frac{1}{3}n^3$ for full elimination $A\backslash b$.

1.5.41 Going from $(5,4,3,2,1)$ to $(1,2,3,4,5)$ requires two exchanges ($1 \leftrightarrow 5$ and $2 \leftrightarrow 4$). Going from $(6,5,4,3,2,1)$ to $(1,2,3,4,5,6)$ requires three exchanges ($1 \leftrightarrow 6$, $2 \leftrightarrow 5$, and $3 \leftrightarrow 4$). In general, $(1,\ldots,4k)$ and $(1,\ldots,4k+1)$ need $2k$ exchanges, because in the $4k+1$ case, the $(2k+1)$th element can stay put. $(1,\ldots,4k+2)$ and $(1,\ldots,4k+3)$ need $2k+1$ exchanges. $n = 100$ and $n = 101$ need 50 exchanges each; $n = 102$ and $n = 103$ need 51.

1.5.43 $P = \begin{bmatrix} 0 & 1 & 0 \\ 0 & 0 & 1 \\ 1 & 0 & 0 \end{bmatrix}$ makes PA upper triangular; $P_1 = \begin{bmatrix} 1 & 0 & 0 \\ 0 & 0 & 1 \\ 0 & 1 & 0 \end{bmatrix}$ and $P_2 = \begin{bmatrix} 0 & 0 & 1 \\ 0 & 1 & 0 \\ 1 & 0 & 0 \end{bmatrix}$ make P_1AP_2 lower triangular. Multiplying a permutation matrix on the right exchanges the *columns* of A.

1.5.45 Any power of a permutation matrix P is itself a permutation matrix. Thus, because there is a finite number (namely $n!$) of permutation matrices of order n, eventually two powers of P must be the same: say $P^r = P^s$ with $s > r$. Then $P^{s-r} = I$, and certainly $r - s \leq n!$.

$P = \begin{bmatrix} P_2 & \\ & P_3 \end{bmatrix}$ is 5 by 5 with $P_2 = \begin{bmatrix} 0 & 1 \\ 1 & 0 \end{bmatrix}$ and $P_3 = \begin{bmatrix} 0 & 1 & 0 \\ 0 & 0 & 1 \\ 1 & 0 & 0 \end{bmatrix}$, and $P^6 = I$. (Note that $P_2 \neq I$ and $P_3^2 \neq I$).

1.5.47 $(I - P)x = Ix - Px$, so take $x = (1,1,\ldots,1)$. Then $Ix = x = Px$, so $I - P$ is singular.

Problem Set 1.3, page 52

1.6.1 $A_1^{-1} = \begin{bmatrix} 0 & \frac{1}{3} \\ \frac{1}{2} & 0 \end{bmatrix}$; $A_2^{-1} = \begin{bmatrix} \frac{1}{2} & 0 \\ -1 & \frac{1}{2} \end{bmatrix}$; $A_3^{-1} = \begin{bmatrix} \cos\theta & \sin\theta \\ -\sin\theta & \cos\theta \end{bmatrix}$.

1.6.3 $A^{-1} = BC^{-1}$; $A^{-1} = U^{-1}L^{-1}P$.

1.6.5 By associativity, $A(AB) = (A^2)B = I$, so $A^{-1} = AB$.

1.6.7 Each of $A = \begin{bmatrix} \frac{\sqrt{3}}{2} & \frac{1}{2} \\ \frac{1}{2} & -\frac{\sqrt{3}}{2} \end{bmatrix}$, $A = \begin{bmatrix} -\frac{\sqrt{3}}{2} & \frac{1}{2} \\ \frac{1}{2} & \frac{\sqrt{3}}{2} \end{bmatrix}$, and $A = \begin{bmatrix} 0 & 1 \\ 1 & 0 \end{bmatrix}$ has the property that $A^2 = I$.

1.6.9 If row 3 of A^{-1} were (a,b,c,d), then $A^{-1}A = I$ gives $2a = 0$, $a + 3b = 0$, $4a + 8b = 1$ (along with another equation). But already this system has no solution.

1.6.11 (a) $\begin{bmatrix} 1 & 0 \\ 0 & 1 \end{bmatrix} + \begin{bmatrix} -1 & 0 \\ 0 & -1 \end{bmatrix} = \begin{bmatrix} 0 & 0 \\ 0 & 0 \end{bmatrix}$

(b) $\begin{bmatrix} 1 & 0 \\ 0 & 0 \end{bmatrix} + \begin{bmatrix} 0 & 0 \\ 0 & 1 \end{bmatrix} = \begin{bmatrix} 1 & 0 \\ 0 & 1 \end{bmatrix}$.

(c) $\begin{bmatrix} 1 & 0 \\ 0 & 1 \end{bmatrix} + \begin{bmatrix} 0 & 1 \\ -1 & 0 \end{bmatrix} = \begin{bmatrix} 1 & 1 \\ -1 & 1 \end{bmatrix}$; $(B^{-1} + A^{-1})^{-1} = B(A+B)^{-1}A$.

1.6.13 $A^{\mathrm{T}}B = 8$; $B^{\mathrm{T}}A = 8$; $AB^{\mathrm{T}} = \begin{bmatrix} 6 & 6 \\ 2 & 2 \end{bmatrix}$; $BA^{\mathrm{T}} = \begin{bmatrix} 6 & 2 \\ 6 & 2 \end{bmatrix}$.

1.6.15 (a) There are $n(n+1)/2$ entries on and above the diagonal, each of which can be chosen independently.

(b) There $n(n-1)/2$ entries above the diagonal.

1.6.17 (a) $A = L_2D_2U_2 = L_1D_1U_1 \Leftrightarrow L_1^{-1}L_2D_2U_2U_2^{-1} = L_1^{-1}L_1D_1U_1U_2^{-1} \Leftrightarrow L_1^{-1}L_2D_2 = D_1U_1U_2^{-1}$. The inverse of a lower (upper) triangular matrix is still lower (upper) triangular. Multiplying lower (upper) triangular matrices gives a lower (upper) triangular matrix.

(b) The main diagonals of $L_1^{-1}L_2D_2$ and $D_1U_1U_2^{-1}$ are the same as those of D_2 and D_1 respectively. $L_1^{-1}L_2D_2 = D_1U_1U_2^{-1}$, so we have $D_1 = D_2$. By comparing the off-diagonals of $L_1^{-1}L_2D_2 = D_1U_1U_2^{-1}$, both matrices must be diagonal. $L_1^{-1}L_2D_2 = D_2$, $D_1U_1U_2^{-1} = D_1$ and D_1 is invertible, so $L_1^{-1}L_2 = I$ and $U_1U_2^{-1} = I$. Thus, $L_1 = L_2$ and $U_1 = U_2$.

1.6.19 $\begin{bmatrix} 1 & 0 & 0 \\ 3 & 1 & 0 \\ 5 & 1 & 1 \end{bmatrix}\begin{bmatrix} 1 & 0 & 0 \\ 0 & 3 & 0 \\ 0 & 0 & 2 \end{bmatrix}\begin{bmatrix} 1 & 3 & 5 \\ 0 & 1 & 1 \\ 0 & 0 & 1 \end{bmatrix} = \begin{bmatrix} 1 & 3 & 5 \\ 3 & 12 & 18 \\ 5 & 18 & 30 \end{bmatrix}$;

$\begin{bmatrix} 1 & 0 \\ b/a & 1 \end{bmatrix}\begin{bmatrix} a & 0 \\ 0 & d-(b^2/a) \end{bmatrix}\begin{bmatrix} 1 & b/a \\ 0 & 1 \end{bmatrix} = \begin{bmatrix} a & b \\ b & d \end{bmatrix}$.

1.6.21 From $B(I-AB) = (I-BA)B$ we get $(I-BA)^{-1} = B(I-AB)^{-1}B^{-1}$, an explicit inverse provided B and $I-AB$ are invertible.

Another approach: If $I - BA$ is not invertible, then $BAx = x$ for some nonzero x. Therefore $ABAx = Ax$, or $ABy = y$, and $I - AB$ could not be invertible. (Note that $y = Ax$ is nonzero from $BAx = x$.)

1.6.23 $\begin{bmatrix} x \\ y \end{bmatrix} = \begin{bmatrix} .5 \\ -.2 \end{bmatrix}$, $\begin{bmatrix} t \\ z \end{bmatrix} = \begin{bmatrix} -.2 \\ .1 \end{bmatrix}$ so $A^{-1} = \frac{1}{10}\begin{bmatrix} 5 & -2 \\ -2 & 1 \end{bmatrix}$.

1.6.25 (a) In $Ax = (1,0,0)$, equation 1 + equation 2 − equation 3 is $0 = 1$

(b) The right sides must satisfy $b_1 + b_2 = b_3$

(c) Row 3 becomes a row of zeros — there is no third pivot.

1.6.27 If B exchanges rows 1 and 2 of A, then B^{-1} exchanges *columns* 1 and 2 of A^{-1}.

1.6.29 If A has a column of zeros, so does BA. So $BA = I$ is impossible. There is no A^{-1}.

1.6.31 $\begin{bmatrix} 1 & & \\ & 1 & \\ & -1 & 1 \end{bmatrix}\begin{bmatrix} 1 & & \\ & 1 & \\ -1 & & 1 \end{bmatrix}\begin{bmatrix} 1 & & \\ -1 & 1 & \\ & & 1 \end{bmatrix} = \begin{bmatrix} 1 & & \\ -1 & 1 & \\ 0 & -1 & 1 \end{bmatrix} = E$; then $\begin{bmatrix} 1 & & \\ 1 & 1 & \\ 1 & 1 & 1 \end{bmatrix}$ is $L = E^{-1}$ after reversing the order of these three elementary matrices and changing -1 to 1.

1.6.33 $A * \text{ones}(4,1)$ is the zero vector, so A cannot be invertible.

1.6.35 $\begin{bmatrix} 1 & 3 & 1 & 0 \\ 2 & 7 & 0 & 1 \end{bmatrix} \to \begin{bmatrix} 1 & 3 & 1 & 0 \\ 0 & 1 & -2 & 1 \end{bmatrix} \to \begin{bmatrix} 1 & 0 & 7 & -3 \\ 0 & 1 & -2 & 1 \end{bmatrix} = \begin{bmatrix} I & A^{-1} \end{bmatrix}$;

$\begin{bmatrix} 1 & 4 & 1 & 0 \\ 3 & 9 & 0 & 1 \end{bmatrix} \to \begin{bmatrix} 1 & 4 & 1 & 0 \\ 0 & -3 & -3 & 1 \end{bmatrix} \to \begin{bmatrix} 1 & 0 & -3 & \frac{4}{3} \\ 0 & 1 & 1 & -\frac{1}{3} \end{bmatrix} = \begin{bmatrix} I & A^{-1} \end{bmatrix}$.

1.6.37 $\begin{bmatrix} 1 & a & b & 1 & 0 & 0 \\ 0 & 1 & c & 0 & 1 & 0 \\ 0 & 0 & 1 & 0 & 0 & 1 \end{bmatrix} \to \begin{bmatrix} 1 & a & 0 & 1 & 0 & -b \\ 0 & 1 & 0 & 0 & 1 & -c \\ 0 & 0 & 1 & 0 & 0 & 1 \end{bmatrix} \to \begin{bmatrix} 1 & 0 & 0 & 1 & -a & ac-b \\ 0 & 1 & 0 & 0 & 1 & -c \\ 0 & 0 & 1 & 0 & 0 & 1 \end{bmatrix}$.

1.6.39 $\begin{bmatrix} 2 & 2 & 0 & 1 \\ 0 & 2 & 1 & 0 \end{bmatrix} \to \begin{bmatrix} 2 & 0 & -1 & 1 \\ 0 & 2 & 1 & 0 \end{bmatrix} \to \begin{bmatrix} 1 & 0 & -\frac{1}{2} & \frac{1}{2} \\ 0 & 1 & \frac{1}{2} & 0 \end{bmatrix} = \begin{bmatrix} I & A^{-1} \end{bmatrix}$.

1.6.41 Not invertible for $c = 7$ (equal columns), $c = 2$ (equal rows), or $c = 0$ (zero column).

1.6.43 $\begin{bmatrix} 1 & -1 & 1 & -1 & 1 & & & \\ & 1 & -1 & 1 & & 1 & & \\ & & 1 & -1 & & & 1 & \\ & & & 1 & & & & 1 \end{bmatrix} \to \begin{bmatrix} 1 & & & & 1 & 1 & & \\ & 1 & & & & 1 & 1 & \\ & & 1 & & & & 1 & 1 \\ & & & 1 & & & & 1 \end{bmatrix} = \begin{bmatrix} I & A^{-1} \end{bmatrix}$. The 5 by 5 A^{-1} also has 1's on the diagonal and superdiagonal and 0's elsewhere.

1.6.45 $\begin{bmatrix} I & 0 \\ -C & I \end{bmatrix}$, $\begin{bmatrix} A^{-1} & 0 \\ -D^{-1}CA^{-1} & D^{-1} \end{bmatrix}$, and $\begin{bmatrix} -D & I \\ I & 0 \end{bmatrix}$.

1.6.47 For $Ax = b$ with $A = \text{ones}(4,4)$ (a singular matrix) and $b = \text{ones}(4,1)$, $A\backslash b$ will pick $x = (1,0,0,0)$ and $\text{pinv}(A) * b$ will pick the shortest solution $x = \frac{1}{4}(1,1,1,1)$.

1.6.49 For $A = \begin{bmatrix} 1 & 0 \\ 9 & 3 \end{bmatrix}$, $A^{\mathrm{T}} = \begin{bmatrix} 1 & 9 \\ 0 & 3 \end{bmatrix}$, $A^{-1} = \begin{bmatrix} 1 & 0 \\ -3 & \frac{1}{3} \end{bmatrix}$, $(A^{-1})^{\mathrm{T}} = (A^{\mathrm{T}})^{-1} = \begin{bmatrix} 1 & -3 \\ 0 & \frac{1}{3} \end{bmatrix}$.

For $A = \begin{bmatrix} 1 & c \\ c & 0 \end{bmatrix}$, $A^{\mathrm{T}} = A$ and then $A^{-1} = \frac{1}{c^2}\begin{bmatrix} 0 & c \\ c & -1 \end{bmatrix} = (A^{-1})^{\mathrm{T}} = (A^{\mathrm{T}})^{-1}$.

1.6.51 (a) $\left((AB)^{-1}\right)^{\mathrm{T}} = (B^{-1}A^{-1})^{\mathrm{T}} = (A^{-1})^{\mathrm{T}}(B^{-1})^{\mathrm{T}}$

(b) $(U^{-1})^{\mathrm{T}}$ is *lower* triangular.

1.6.53 (a) $x^{\mathrm{T}}Ay = a_{22} = 5$ (b) $x^{\mathrm{T}}A = \begin{bmatrix} 4 & 5 & 6 \end{bmatrix}$ (c) $Ay = \begin{bmatrix} 2 \\ 5 \end{bmatrix}$

1.6.55 $(Px)^{\mathrm{T}}(Py) = x^{\mathrm{T}}P^{\mathrm{T}}Py = x^{\mathrm{T}}y$ because $P^{\mathrm{T}}P = I$; usually $Px \cdot y = x \cdot P^{\mathrm{T}}y \neq x \cdot Py$:

$\begin{bmatrix} 0 & 1 & 0 \\ 0 & 0 & 1 \\ 1 & 0 & 0 \end{bmatrix}\begin{bmatrix} 1 \\ 2 \\ 3 \end{bmatrix} \cdot \begin{bmatrix} 1 \\ 1 \\ 2 \end{bmatrix} = 7$, but $\begin{bmatrix} 1 \\ 2 \\ 3 \end{bmatrix} \cdot \begin{bmatrix} 0 & 1 & 0 \\ 0 & 0 & 1 \\ 1 & 0 & 0 \end{bmatrix}\begin{bmatrix} 1 \\ 1 \\ 2 \end{bmatrix} = 8$.

1.6.57 PAP^{T} recovers the symmetry.

1.6.59 (a) The transpose of $R^{\mathrm{T}}AR$ is $R^{\mathrm{T}}A^{\mathrm{T}}R^{\mathrm{TT}} = R^{\mathrm{T}}AR$, an n by n matrix.

(b) $(R^{\mathrm{T}}R)_{jj}$ = (column j of R) $\cdot$ (column j of R) = (length of column j)^2

1.6.61 (a) Total currents are $A^{\mathrm{T}}y = \begin{bmatrix} 1 & 0 & 1 \\ -1 & 1 & 0 \\ 0 & -1 & -1 \end{bmatrix}\begin{bmatrix} y_{\mathrm{BC}} \\ y_{\mathrm{CS}} \\ y_{\mathrm{BS}} \end{bmatrix} = \begin{bmatrix} y_{\mathrm{BC}} + y_{\mathrm{BS}} \\ -y_{\mathrm{BC}} + y_{\mathrm{CS}} \\ -y_{\mathrm{CS}} - y_{\mathrm{BS}} \end{bmatrix}$.

(b) $(Ax)^{\mathrm{T}}y = x^{\mathrm{T}}(A^{\mathrm{T}}y) = x_{\mathrm{B}}y_{\mathrm{BC}} + x_{\mathrm{B}}y_{\mathrm{BS}} - x_{\mathrm{C}}y_{\mathrm{BC}} + x_{\mathrm{C}}y_{\mathrm{CS}} - x_{\mathrm{S}}y_{\mathrm{CS}} - x_{\mathrm{S}}y_{\mathrm{BS}}$.

1.6.63 $Ax \cdot y$ is the *cost* of inputs while $x \cdot A^{\mathrm{T}}y$ is the value of *outputs*.

1.6.65 These are groups: Lower triangular with diagonal 1's, diagonal invertible D, and permutations P. Two more: Even permutations, all nonsingular matrices.

1.6.67 Reordering the rows and/or columns of $\left[\begin{smallmatrix} a & b \\ c & d \end{smallmatrix}\right]$ will move entry a, not giving $\left[\begin{smallmatrix} a & c \\ b & d \end{smallmatrix}\right]$.

1.6.69 A random matrix is almost sure to be invertible.

1.6.71 The $-1, 2, -1$ matrix in Equation 1.7.6 has $A = LDL^{\mathrm{T}}$ with $\ell_{i,i-1} = 1 - \frac{1}{i}$.

Problem Set 1.7, page 63

1.7.1
$$\begin{bmatrix} 2 & -1 & & \\ -1 & 2 & -1 & \\ & -1 & 2 & -1 \\ & & -1 & 2 \end{bmatrix} = \begin{bmatrix} 1 & & & \\ -\frac{1}{2} & 1 & & \\ & -\frac{2}{3} & 1 & \\ & & -\frac{3}{4} & 1 \end{bmatrix} \begin{bmatrix} 2 & & & \\ & \frac{3}{2} & & \\ & & \frac{4}{3} & \\ & & & \frac{5}{4} \end{bmatrix} \begin{bmatrix} 1 & -\frac{1}{2} & & \\ & 1 & -\frac{2}{3} & \\ & & 1 & -\frac{3}{4} \\ & & & 1 \end{bmatrix} = LDL^{\mathrm{T}}.$$
Determinant is $\frac{2}{1} \cdot \frac{3}{2} \cdot \frac{4}{3} \cdot \frac{5}{4} = 5$.

1.7.3 $A_0 = \begin{bmatrix} 1 & -1 & & & \\ -1 & 2 & -1 & & \\ & -1 & 2 & -1 & \\ & & -1 & 2 & -1 \\ & & & -1 & 1 \end{bmatrix}$. Each row adds to 1, so $A_0 \begin{bmatrix} c \\ c \\ c \\ c \\ c \end{bmatrix} = \begin{bmatrix} 0 \\ 0 \\ 0 \\ 0 \\ 0 \end{bmatrix}$.

1.7.5 $(u_1, u_2, u_3) = \left(\frac{1}{8}\pi^2, 0, -\frac{1}{8}\pi^2\right) \approx (1.234, 0, -1.234)$ instead of the true values $(1, 0, -1)$.

1.7.7 The true inverse is $H^{-1} = \begin{bmatrix} 9 & -36 & 30 \\ -36 & 192 & -180 \\ 30 & -180 & 180 \end{bmatrix}$. Rounding gives $H \approx \begin{bmatrix} 1 & .5 & .333 \\ .5 & .333 & .25 \\ .333 & .25 & .2 \end{bmatrix} \Rightarrow$

$$H^{-1} \approx \begin{bmatrix} 9.671 & -39.508 & 33.284 \\ -39.508 & 210.186 & -196.951 \\ 33.284 & -196.951 & 195.772 \end{bmatrix}.$$

1.7.9 The actual solution is
$x = (100, -4950, 79200, -600600, 2522520, -6306300, 9609600, -8751600, 4375800, -923780)$.
If we replace $h_{12} = 0.5$ with 0.4999, we get
$x \approx (67, -3311, 52977, -401739, 1687304, -4218261, 6427826, -5853913, 2926957, -617913)$.
The 10 by 10 Hilbert matrix is very ill-conditioned.

1.7.11 A large pivot is multiplied by less than one in eliminating each entry below it. An extreme case, with multipliers 1 and pivots $\frac{1}{2}$, $\frac{1}{2}$, and 4, is $A = \begin{bmatrix} \frac{1}{2} & \frac{1}{2} & 1 \\ -\frac{1}{2} & 0 & 1 \\ -\frac{1}{2} & -1 & 1 \end{bmatrix}$.

Chapter 2 Vector Spaces

Problem Set 2.1, page 73

2.1.1 (a) The set of all (u, v), where u and v are ratios p/q of integers.

(b) The set of all (u, v), where $u = 0$ or $v = 0$.

2.1.3 $\boldsymbol{C}(A)$ is the x-axis; $\boldsymbol{N}(A)$ is the line through $(1, 1)$; $\boldsymbol{C}(B)$ is $\mathbf{R}^2$; $\boldsymbol{N}(B)$ is the line through $(-2, 1, 0)$; $\boldsymbol{C}(C)$ is the point $(0, 0)$ in $\mathbf{R}^2$; the nullspace $\boldsymbol{N}(C)$ is $\mathbf{R}^3$.

2.1.5 (a) Rules 7 and 8 are broken.

(b) 1: $xy = yx$ (which is $x + y = y + x$ with the new rule for addition)
2: $x\,(yz) = (xy)\,z$
3: $x\,(1) = x$ for all x (hence, the "zero vector" is 1.)
4: $x\left(\frac{1}{x}\right) = 1$ (the zero vector) for all $x > 0$
5: $x^1 = x$
6: $x^{c_1 c_2} = (x^{c_2})^{c_1}$. Note that the multiplication of the scalars c_1 and c_2 is "normal" multiplication.
7: $(xy)^c = x^c y^c$
8: $x^{c_1 + c_2} = x^{c_1} x^{c_2}$. Note that the addition of the scalars c_1 and c_2 is "normal" addition.

(c) Rules 1, 2, and 8 are not satisfied.

2.1.7 (b), (d), (e) are subspaces. (a) and (f) fail to be subspaces because they are not closed under addition. (c) fails to be a subspace because it is not closed under scalar multiplication (the negative of any decreasing sequence is an increasing sequence).

2.1.9 The sum of two nonsingular matrices may be singular; for example, $A + (-A) = 0$. Furthermore, the sum of two singular matrices may be nonsingular.

2.1.11 (a) One possibility: The matrices cA form a subspace not containing B.

(b) Yes: the subspace must contain $A - B = I$.

(c) One possibility: the subspace of matrices whose main diagonals consist entirely of 0's.

2.1.13 If $(f + g)\,(x)$ is the usual $f\,(g\,(x))$, then $(g + f)\,x = g\,(f\,(x))$ which is different. In Rule 2, both sides are $f\,(g\,(h\,(x)))$. Rule 4 is broken because there might be no inverse function $f^{-1}\,(x)$ such that $f\left(f^{-1}\,(x)\right) = x$. If the inverse function exists it will be the vector $-f$.

2.1.15 The sum of $(4, 0, 0)$ and $(0, 4, 0)$ is not on the plane; it has $x + y - 2z = 8$.

2.1.17 (a) The subspaces of $\mathbf{R}^2$ are $\mathbf{R}^2$ itself, lines through $(0, 0)$, and the point $(0, 0)$.

(b) The subspaces of $\mathbf{R}^4$ are $\mathbf{R}^4$ itself, three-dimensional "planes" $n \cdot v = 0$, two-dimensional subspaces ($n_1 \cdot v = 0$ and $n_2 \cdot v = 0$), one-dimensional lines through $(0, 0, 0, 0)$, and $(0, 0, 0, 0)$ alone.

2.1.19 The smallest subspace containing $\mathbf{P}$ and $\mathbf{L}$ is either $\mathbf{P}$ (if $\mathbf{L}$ lies on $\mathbf{P}$) or $\mathbf{R}^3$ (otherwise).

2.1.21 The column space of A is the x-axis, that is, the set of all vectors $(x,0,0)$. The column space of B is the x-y plane, that is, the set of all vectors $(x,y,0)$. The column space of C is the line of vectors $(x,2x,0)$.

2.1.23 A combination of the columns of C is also a combination of the columns of A. A and C have the same column space; B has a different column space.

2.1.25 The extra column b enlarges the column space unless b is *already in* that space: $\begin{bmatrix} A & b \end{bmatrix} = \begin{bmatrix} 1 & 0 & 1 \\ 0 & 0 & 1 \end{bmatrix}$ has a larger column space because $Ax = b$ has no solution. $\begin{bmatrix} 1 & 0 & 1 \\ 0 & 1 & 1 \end{bmatrix}$ has the same column space because $Ax = b$ has a solution; b is already in the column space.

2.1.27 Column space $= \mathbf{R}^8$. Every b is a combination of the columns, since $Ax = b$ is solvable.

2.1.29 The column spaces of $A = \begin{bmatrix} 1 & 1 & 0 \\ 1 & 0 & 0 \\ 0 & 1 & 0 \end{bmatrix}$ and $B = \begin{bmatrix} 1 & 1 & 2 \\ 1 & 0 & 1 \\ 0 & 1 & 1 \end{bmatrix}$ contain $(1,1,0)$ and $(1,0,1)$, but not $(1,1,1)$. The column space of $C = \begin{bmatrix} 1 & 2 & 0 \\ 2 & 4 & 0 \\ 3 & 6 & 0 \end{bmatrix}$ is the line $(c,2c,3c)$.

2.1.31 $\mathbf{R}^2$ contains vectors with *two* components — they don't belong to $\mathbf{R}^3$.

Problem Set 2.2, page 85

2.2.1 $x+y+z=1$, $x+y+z=0$. Changing 1 to 0, $(x,y,z) = c(-1,1,0) + d(-1,0,1)$.

2.2.3 Echelon form $U = \begin{bmatrix} 0 & 1 & 0 & 3 \\ 0 & 0 & 0 & 0 \end{bmatrix}$; free variables x_1, x_3, and x_4; special solutions $(1,0,0,0)$, $(0,0,1,0)$, and $(0,-3,0,1)$. The system is consistent when $b_2 = 2b_1$. Complete solution $(0,b_1,0,0)$ plus any combination of special solutions.

2.2.5 The first system has complete solutions $\begin{bmatrix} u \\ v \\ w \end{bmatrix} = \begin{bmatrix} -2v-3 \\ v \\ 2 \end{bmatrix} = v\begin{bmatrix} -2 \\ 1 \\ 0 \end{bmatrix} + \begin{bmatrix} -3 \\ 0 \\ 2 \end{bmatrix}$. The second system has no solution.

2.2.7 $c = 7$ allows $u = 1$, $v = 1$, $w = 0$. The column space is a plane.

2.2.9 (a) $x = x_2\begin{bmatrix} -2 \\ 1 \\ 0 \\ 0 \end{bmatrix} + x_4\begin{bmatrix} 2 \\ 0 \\ -2 \\ 1 \end{bmatrix}$ for any x_2, x_4. Row-reduced $R = \begin{bmatrix} 1 & 2 & 0 & -2 \\ 0 & 0 & 1 & 2 \\ 0 & 0 & 0 & 0 \end{bmatrix}$.

(b) Complete solution $x = \begin{bmatrix} a-3b \\ 0 \\ b \\ 0 \end{bmatrix} + x_2\begin{bmatrix} -2 \\ 1 \\ 0 \\ 0 \end{bmatrix} + x_4\begin{bmatrix} 2 \\ 0 \\ -2 \\ 1 \end{bmatrix}$ for any x_2, x_4.

2.2.11 The nullspace of $\begin{bmatrix} 1 & 1 \\ 1 & 1 \end{bmatrix} \begin{bmatrix} x_1 \\ x_2 \end{bmatrix} = \begin{bmatrix} 1 \\ 0 \end{bmatrix}$ is the line through $(-1, 1)$, but there is no solution. Any $b = \begin{bmatrix} c \\ c \end{bmatrix}$ has many particular solutions to $Ax_p = b$.

2.2.13 (a) $R = \begin{bmatrix} 1 & 1 & 1 & 1 \\ 0 & 0 & 0 & 0 \\ 0 & 0 & 0 & 0 \end{bmatrix}$; rank is 1.

(b) $R = \begin{bmatrix} 1 & 0 & 1 & 0 \\ 0 & 1 & 0 & 1 \\ 0 & 0 & 0 & 0 \\ 0 & 0 & 0 & 0 \end{bmatrix}$; rank is 2.

(c) $R = \begin{bmatrix} 1 & -1 & 1 & -1 \\ 0 & 0 & 0 & 0 \\ 0 & 0 & 0 & 0 \end{bmatrix}$; rank is 1.

2.2.15 A nullspace matrix $N = \begin{bmatrix} -F \\ I \end{bmatrix}$ is n by $n - r$.

2.2.17 This is not necessarily true if the echelon form is not reduced. If $A_1 = \begin{bmatrix} 1 & 0 & 0 \\ 1 & 0 & 0 \end{bmatrix}$ and $A_2 = \begin{bmatrix} 1 & 0 & 0 \\ -1 & 0 & 0 \end{bmatrix}$, then $A_1 + A_2 = R_1 + R_2 = \begin{bmatrix} 2 & 0 & 0 \\ 0 & 0 & 0 \end{bmatrix}$, but $A_1 \neq R_1$ and $A_2 \neq R_2$.
If A_1 and A_2 are in *reduced* row echelon form, then necessarily $A_1 = R_1$ and $A_2 = R_2$; also $A_1 \neq A_2$. This is true. We must have one of $A_1 = 0$ or $A_2 = 0$, otherwise A_1 and A_2 both have nonzero first pivots and $A_1 + A_2$ reduces further.

2.2.19 The special solutions are the columns of $N = \begin{bmatrix} -2 & -3 \\ -4 & -5 \\ 1 & 0 \\ 0 & 1 \end{bmatrix}$ and $N = \begin{bmatrix} 1 & 0 \\ 0 & -2 \\ 0 & 1 \end{bmatrix}$.

2.2.21 The r pivot columns of A form an m by r submatrix of rank r. So that matrix A^* has r independent pivot rows, giving an r by r invertible submatrix of A. (The pivot rows of A^* and A are the same, since elimination is done in the same order — we just don't see for A^* the "free" columns of zeros that appear for A.)

2.2.23 $(uv^{\mathrm{T}})(wz^{\mathrm{T}}) = u(v^{\mathrm{T}}w)z^{\mathrm{T}}$ has rank 1 unless $v^{\mathrm{T}}w = 0$.

2.2.25 We are given $AB = I$ which has rank n. Then $\operatorname{rank}(AB) = n \leq \operatorname{rank}(A)$, but A is n by n and so has rank at most n. Thus $n \leq \operatorname{rank}(A) \leq n$ forces $\operatorname{rank}(A) = n$.

2.2.27 If $R = EA$ and the same $R = E^*B$, then $B = (E^*)^{-1} EA$. (To get B, reduce A to R and then invert steps back to B.) B is an *invertible* matrix times A, when they share the same R.

2.2.29 Since R starts with r independent rows, R^{T} starts with r independent columns (and then zeros). So *its* reduced echelon form is $\begin{bmatrix} I & 0 \\ 0 & 0 \end{bmatrix}$ where I is r by r.

2.2.31 For $A = \begin{bmatrix} 1 & 1 & 2 & 2 \\ 2 & 2 & 4 & 4 \\ 1 & c & 2 & 2 \end{bmatrix}$:

If $c = 1$, $R = \begin{bmatrix} 1 & 1 & 2 & 2 \\ 0 & 0 & 0 & 0 \\ 0 & 0 & 0 & 0 \end{bmatrix}$ has x_2, x_3, and x_4 free.

If $c \neq 1$, $R = \begin{bmatrix} 1 & 0 & 2 & 2 \\ 0 & 1 & 0 & 0 \\ 0 & 0 & 0 & 0 \end{bmatrix}$ has x_3 and x_4 free.

Special solutions in $N = \begin{bmatrix} -1 & -2 & -2 \\ 1 & 0 & 0 \\ 0 & 1 & 0 \\ 0 & 0 & 1 \end{bmatrix}$ $(c = 1)$ and $N = \begin{bmatrix} -2 & -2 \\ 0 & 0 \\ 1 & 0 \\ 0 & 1 \end{bmatrix}$ $(c \neq 1)$.

For $A = \begin{bmatrix} 1-c & 2 \\ 0 & 2-c \end{bmatrix}$:

If $c = 1$, $R = \begin{bmatrix} 0 & 1 \\ 0 & 0 \end{bmatrix}$ has x_1 free; if $c = 2$, $R = \begin{bmatrix} 1 & -2 \\ 0 & 0 \end{bmatrix}$ has x_2 free. $R = I$ if $c \neq 1$ or 2.

Special solutions in $N = \begin{bmatrix} 1 \\ 0 \end{bmatrix}$ $(c = 1)$ or $N = \begin{bmatrix} 2 \\ 1 \end{bmatrix}$ $(c = 2)$. If $c \neq 1$ or 2, N is the empty 2 by 0 matrix.

2.2.33 $\begin{bmatrix} 1 & 3 & 3 & 1 \\ 2 & 6 & 9 & 5 \\ -1 & -3 & 3 & 5 \end{bmatrix} \rightarrow \begin{bmatrix} 1 & 3 & 3 & 1 \\ 0 & 0 & 3 & 3 \\ 0 & 0 & 0 & 0 \end{bmatrix}$, so back-substitution gives $3z = 3$ or $z = 1$, $x + 3y + 3 = 1$ or $x + 3y = -2$. $y = 0$ gives the particular solution $\begin{bmatrix} -2 \\ 0 \\ 1 \end{bmatrix}$, so $x_{\text{complete}} = \begin{bmatrix} -2 \\ 0 \\ 1 \end{bmatrix} + x_2 \begin{bmatrix} -3 \\ 1 \\ 0 \end{bmatrix}$.

$\begin{bmatrix} 1 & 3 & 1 & 2 & 1 \\ 2 & 6 & 4 & 8 & 3 \\ 0 & 0 & 2 & 4 & 1 \end{bmatrix} \rightarrow \begin{bmatrix} 1 & 3 & 1 & 2 & 1 \\ 0 & 0 & 2 & 4 & 1 \\ 0 & 0 & 0 & 0 & 0 \end{bmatrix}$. Back-substitution gives $z + 2t = \frac{1}{2}$, $x + 3y = \frac{1}{2}$ If $y = 0$ and $t = 0$, we get $x_p = \begin{bmatrix} \frac{1}{2} \\ 0 \\ \frac{1}{2} \\ 0 \end{bmatrix}$, so $x_{\text{complete}} = \begin{bmatrix} \frac{1}{2} \\ 0 \\ \frac{1}{2} \\ 0 \end{bmatrix} + x_2 \begin{bmatrix} -3 \\ 1 \\ 0 \\ 0 \end{bmatrix} + x_4 \begin{bmatrix} 0 \\ 0 \\ -2 \\ 1 \end{bmatrix}$.

2.2.35 (a) $\begin{bmatrix} 1 & 2 & b_1 \\ 2 & 4 & b_2 \\ 2 & 5 & b_3 \\ 3 & 9 & b_4 \end{bmatrix} \rightarrow \begin{bmatrix} 1 & 2 & b_1 \\ 0 & 1 & b_3 - 2b_1 \\ 0 & 3 & b_4 - 3b_1 \\ 0 & 0 & b_2 - 2b_1 \end{bmatrix} \rightarrow \begin{bmatrix} 1 & 2 & b_1 \\ 0 & 1 & b_3 - 2b_1 \\ 0 & 0 & 3b_1 - 3b_3 + b_4 \\ 0 & 0 & b_2 - 2b_1 \end{bmatrix}$, so the system is solvable if $b_2 = 2b_1$ and $3b_1 - 3b_3 + b_4 = 0$. Then $x = \begin{bmatrix} 5b_1 - 2b_3 \\ b_3 - 2b_1 \end{bmatrix}$ (no free variables).

(b) $\begin{bmatrix} 1 & 2 & 3 & b_1 \\ 2 & 4 & 6 & b_2 \\ 2 & 5 & 7 & b_3 \\ 3 & 9 & 12 & b_4 \end{bmatrix} \to \begin{bmatrix} 1 & 2 & 3 & b_1 \\ 0 & 0 & 0 & b_2 - 2b_1 \\ 0 & 1 & 1 & b_3 - 2b_1 \\ 0 & 3 & 3 & b_4 - 3b_1 \end{bmatrix} \to \begin{bmatrix} 1 & 2 & 3 & b_1 \\ 0 & 1 & 1 & b_3 - 2b_1 \\ 0 & 0 & 0 & 3b_1 - 3b_3 + b_4 \\ 0 & 0 & 0 & b_2 - 2b_1 \end{bmatrix}$ is solvable if

$b_2 = 2b_1$ and $3b_1 - 3b_3 + b_4 = 0$. Then $x = \begin{bmatrix} 5b_1 - 2b_3 \\ b_3 - 2b_1 \\ 0 \end{bmatrix} + x_3 \begin{bmatrix} -1 \\ -1 \\ 1 \end{bmatrix}$.

2.2.37 A 1 by 3 system has at least two free variables.

2.2.39 (a) The particular solution x_p is always multiplied by 1.

(b) Any solution can be x_p.

(c) $\begin{bmatrix} 3 & 3 \\ 3 & 3 \end{bmatrix}\begin{bmatrix} x \\ y \end{bmatrix} = \begin{bmatrix} 6 \\ 6 \end{bmatrix}$. Then $\begin{bmatrix} 1 \\ 1 \end{bmatrix}$ is shorter (length $\sqrt{2}$) than $\begin{bmatrix} 2 \\ 0 \end{bmatrix}$.

(d) The "homogeneous" solution in the nullspace is $x_n = 0$ when A is invertible.

2.2.41 Multiply x_p by 2, same x_n; $\begin{bmatrix} x \\ X \end{bmatrix}_p$ is $\begin{bmatrix} x_p \\ 0 \end{bmatrix}$, special solutions also include the columns of $\begin{bmatrix} -I \\ I \end{bmatrix}$; x_p and the special solutions are not changed.

2.2.43 For A, $q = 3$ gives rank 1; every other q gives rank 2; rank 3 is impossible. For B, $q = 6$ gives rank 1; every other q gives rank 2; rank 3 is obviously impossible since B has only two rows.

2.2.45 (a) $r < m$, always $r \leq n$. (b) $r = m, r < n$

(c) $r < m, r = n$ (d) $r = m = n$.

2.2.47 $R = \begin{bmatrix} 1 & 0 & 0 & 0 \\ 0 & 0 & 1 & 0 \\ 0 & 0 & 0 & 0 \end{bmatrix}$ and $x_n = \begin{bmatrix} 0 \\ 1 \\ 0 \end{bmatrix}$; $\begin{bmatrix} 1 & 0 & 0 & -1 \\ 0 & 0 & 1 & 2 \\ 0 & 0 & 0 & 5 \end{bmatrix}$ has no solution because of row 3.

2.2.49 (a) $A = \begin{bmatrix} 1 & 1 \\ 0 & 2 \\ 0 & 3 \end{bmatrix}$

(b) B can't exist since two equations in three unknowns can't have exactly one solution.

2.2.51 (a) A has rank $4 - 1 = 3$. The complete solution to $Ax = 0$ is $x = (2, 3, 1, 0)$.

(b) $R = \begin{bmatrix} 1 & 0 & -2 & 0 \\ 0 & 1 & -3 & 0 \\ 0 & 0 & 0 & 1 \end{bmatrix}$ with $-2, -3$ in the free column.

2.2.53 (a) False (b) True (c) True (only n columns) (d) True (only m rows)

2.2.55 $U = \begin{bmatrix} 0 & 1 & 1 & 1 & 1 & 1 & 1 \\ 0 & 0 & 0 & 1 & 1 & 1 & 1 \\ 0 & 0 & 0 & 0 & 1 & 1 & 1 \\ 0 & 0 & 0 & 0 & 0 & 0 & 0 \end{bmatrix}$; $R = \begin{bmatrix} 0 & 1 & 1 & 0 & 0 & 1 & 1 \\ 0 & 0 & 0 & 1 & 0 & 1 & 1 \\ 0 & 0 & 0 & 0 & 1 & 1 & 1 \\ 0 & 0 & 0 & 0 & 0 & 0 & 0 \end{bmatrix}$. Note that this R does not correspond to this U.

2.2.57 If column 1 = column 5, then x_5 is a free variable. Its special solution is $(-1, 0, 0, 0, 1)$.

2.2.59 Column 5 is sure to have no pivot since it is a combination of earlier columns, and x_5 is free. With four pivots in the other columns, the special solution is $(1, 0, 1, 0, 1)$. The nullspace contains all multiples of $(1, 0, 1, 0, 1)$ (a line in $\mathbf{R}^5$).

2.2.61 $A = \begin{bmatrix} 1 & 0 & 0 & -4 \\ 0 & 1 & 0 & -3 \\ 0 & 0 & 1 & -2 \end{bmatrix}$

2.2.63 This construction is impossible: there are two pivot columns and two free variables, but only three columns.

2.2.65 $A = \begin{bmatrix} 0 & 1 \\ 0 & 0 \end{bmatrix}$

2.2.67 R is most likely to be I; R is most likely to be $\begin{bmatrix} & I & \\ 0 & 0 & 0 \end{bmatrix}$.

2.2.69 Any zero rows come after these rows: $R = \begin{bmatrix} 1 & -2 & -3 \end{bmatrix}$, $R = \begin{bmatrix} 1 & 0 & 0 \\ 0 & 1 & 0 \end{bmatrix}$, $R = I$.

Problem Set 2.3, page 98

2.3.1 $\begin{bmatrix} 1 & 1 & 1 \\ 0 & 1 & 1 \\ 0 & 0 & 1 \end{bmatrix} \begin{bmatrix} c_1 \\ c_2 \\ c_3 \end{bmatrix} = 0$ gives $c_3 = c_2 = c_1 = 0$. But $v_1 + v_2 - 4v_3 + v_4 = 0$ (dependent).

2.3.3 If $a = 0$ then column $1 = 0$; if $d = 0$ then b (column 1) $- a$ (column 2) $= 0$; if $f = 0$ then all columns end in zero (all are perpendicular to $(0, 0, 1)$ and lie in the xy plane, so must be dependent).

2.3.5 (a) $\begin{bmatrix} 1 & 2 & 3 \\ 3 & 1 & 2 \\ 2 & 3 & 1 \end{bmatrix} \to \begin{bmatrix} 1 & 2 & 3 \\ 0 & -5 & -7 \\ 0 & -1 & -5 \end{bmatrix} \to \begin{bmatrix} 1 & 2 & 3 \\ 0 & -5 & -7 \\ 0 & 0 & -\frac{18}{5} \end{bmatrix}$ is invertible, so the columns are independent. (We could have used rows.)

(b) $\begin{bmatrix} 1 & 2 & -3 \\ -3 & 1 & 2 \\ 2 & -3 & 1 \end{bmatrix} \to \begin{bmatrix} 1 & 2 & -3 \\ 0 & 7 & -7 \\ 0 & 0 & 0 \end{bmatrix}$ is singular, so the columns are dependent. $A \begin{bmatrix} 1 \\ 1 \\ 1 \end{bmatrix} = \begin{bmatrix} 0 \\ 0 \\ 0 \end{bmatrix}$.

Again, we could have used rows.

2.3.7 The sum $v_1 - v_2 + v_3 = (w_2 - w_3) - (w_1 - w_3) + (w_1 - w_2) = 0$.

2.3.9 (a) The four vectors are the columns of a 3 by 4 matrix A with at least one free variable, so $Ax = 0$.

(b) v_1 and v_2 are dependent if $\begin{bmatrix} v_1 & v_2 \end{bmatrix}$ has rank 0 or 1.

(c) v_1 and $(0, 0, 0)$ are dependent because $0v_1 + c\,(0, 0, 0) = 0$ has a nonzero solution (take any $c \neq 0$).

2.3.11 (a) Line in $\mathbf{R}^3$ (b) Plane in $\mathbf{R}^3$ (c) Plane in $\mathbf{R}^3$ (d) All of $\mathbf{R}^3$

2.3.13 All dimensions are 2. The row spaces of A and U are the same.

2.3.15 $v = \frac{1}{2}(v + w) + \frac{1}{2}(v - w)$ and $w = \frac{1}{2}(v + w) - \frac{1}{2}(v - w)$. The two pairs *span* the same space. They are a basis when v and w are *independent*.

2.3.17 If elimination produces one or more zero rows, the rows of A are linearly dependent. For example, in

Problem 16 $\begin{bmatrix} 1 & 1 & 0 & 0 \\ 1 & 0 & 1 & 0 \\ 0 & 0 & 1 & 1 \\ 0 & 1 & 0 & 1 \end{bmatrix} \to \begin{bmatrix} 1 & 1 & 0 & 0 \\ 0 & -1 & 1 & 0 \\ 0 & 0 & 1 & 1 \\ 0 & 0 & 1 & 1 \end{bmatrix} \to \begin{bmatrix} 1 & 1 & 0 & 0 \\ 0 & -1 & 1 & 0 \\ 0 & 0 & 1 & 1 \\ 0 & 0 & 0 & 0 \end{bmatrix}$.

2.3.19 The n independent vectors span a space of dimension n. They are a *basis* for that space. If they are the columns of A then m is *not less* than n (that is, $m \geq n$).

2.3.21 Any basis for $\mathbf{R}^2$ forms a basis for $\mathbf{C}(U)$. Two possible bases for the row space of U are $\{\text{row } 1, \text{row } 2\}$ and $\{\text{row } 1, \text{row } 1 + \text{row } 2\}$.

2.3.23 If the columns are independent, then A has rank n. If the columns span $\mathbf{R}^m$, then A has rank m. If the columns are basis for $\mathbf{R}^m$, then A is square and $\operatorname{rank}(A) = m = n$.

2.3.25 (a) The only solution is $x = 0$ because *the columns are independent*.

(b) $Ax = b$ is solvable because *the columns span* $\mathbf{R}^5$.

2.3.27 Columns 1 and 2 of each matrix are bases for the (different) column spaces of each matrix. Rows 1 and 2 are bases for the (equal) row spaces; $(1, -1, 1)$ is a basis for the (equal) nullspaces.

2.3.29 $\operatorname{rank}(A) = 2$ if $c = 0$ and $d = 2$; $\operatorname{rank}(B) = 2$ except when $c = \pm d$.

2.3.31 Let $v_1 = (1, 0, 0, 0), \ldots, v_4 = (0, 0, 0, 1)$ be the coordinate vectors. If $\mathbf{W}$ is the line through $(1, 2, 3, 4)$, none of the v's are in $\mathbf{W}$.

2.3.33 (a) If it were not a basis, we could add more independent vectors, which would exceed the given dimension k.

(b) If it were not a basis, we could omit some vectors, leaving less than the given dimension k.

2.3.35 (a) False. There might be no solution.

(b) True. Any seven vectors in $\mathbf{R}^5$ are dependent.

2.3.37 (a) $\left\{ \begin{bmatrix} 1 & 0 & 0 \\ 0 & 0 & 0 \\ 0 & 0 & 0 \end{bmatrix}, \begin{bmatrix} 0 & 0 & 0 \\ 0 & 1 & 0 \\ 0 & 0 & 0 \end{bmatrix}, \begin{bmatrix} 0 & 0 & 0 \\ 0 & 0 & 0 \\ 0 & 0 & 1 \end{bmatrix} \right\}$

(b) The above set plus $\left\{ \begin{bmatrix} 0 & 1 & 0 \\ 1 & 0 & 0 \\ 0 & 0 & 0 \end{bmatrix}, \begin{bmatrix} 0 & 0 & 1 \\ 0 & 0 & 0 \\ 1 & 0 & 0 \end{bmatrix}, \begin{bmatrix} 0 & 0 & 0 \\ 0 & 0 & 1 \\ 0 & 1 & 0 \end{bmatrix} \right\}$.

(c) $\left\{ \begin{bmatrix} 0 & 1 & 0 \\ -1 & 0 & 0 \\ 0 & 0 & 0 \end{bmatrix}, \begin{bmatrix} 0 & 0 & 1 \\ 0 & 0 & 0 \\ -1 & 0 & 0 \end{bmatrix}, \begin{bmatrix} 0 & 0 & 0 \\ 0 & 0 & 1 \\ 0 & -1 & 0 \end{bmatrix} \right\}$

2.3.39 $y(0) = 0$ requires $A + B + C = 0$. One basis is $\{\cos x - \cos 2x, \cos x - \cos 3x\}$.

2.3.41 $\operatorname{span}(\{x, 2x, 3x\}) = \operatorname{span}(x)$ has dimension 1; $\operatorname{span}\left(\{x, 2x, x^2\}\right)$ has dimension 2; $\operatorname{span}\left(\{x, x^2, x^3\}\right)$ has dimension 3.

2.3.43 $\begin{bmatrix} 1 & & \\ & 1 & \\ & & 1 \end{bmatrix} = \begin{bmatrix} & 1 & \\ 1 & & \\ & & 1 \end{bmatrix} - \begin{bmatrix} & 1 & \\ & & 1 \\ 1 & & \end{bmatrix} + \begin{bmatrix} & & 1 \\ & 1 & \\ 1 & & \end{bmatrix} + \begin{bmatrix} 1 & & \\ & & 1 \\ & 1 & \end{bmatrix} - \begin{bmatrix} & & 1 \\ 1 & & \\ & 1 & \end{bmatrix}$.

Check the $(1, 1)$ entry, then $(3, 2)$, then $(3, 3)$, then $(1, 2)$ to show that those five P's are independent. Four conditions on the nine entries make all row sums and column sums equal: row sum $1 =$ row sum $2 =$ row sum $3 =$ column sum $1 =$ column sum 2. column sum 3 is automatically equal, because the sum of all rows is equal to the sum of all columns.

2.3.45 If the 5 by 5 matrix $\begin{bmatrix} A & b \end{bmatrix}$ is invertible, b is not a combination of the columns of A. If $\begin{bmatrix} A & b \end{bmatrix}$ is singular and A has independent columns, b *is* a combination of those columns.

Problem Set 2.4, page 110

2.4.1 False; we know only that the dimensions are equal. The nullspace has a larger dimension than the *left* nullspace, which has dimension $m - r$.

2.4.3 $\boldsymbol{C}(A)$: $r = 2$, $\{(1,0,1)\,(0,1,0)\}$; $\boldsymbol{N}(A)$: $n - r = 2$, $\{(2,-1,1,0),(-1,0,0,1)\}$;
$\boldsymbol{C}(A^{\mathrm{T}})$: $r = 2$, $\{(1,2,0,1),(0,1,1,0)\}$; $\boldsymbol{N}(A^{\mathrm{T}})$: $m - r = 1$, $\{(-1,0,1)\}$;
$\boldsymbol{C}(U)$: $\{(1,0,0),(0,1,0)\}$; $\boldsymbol{N}(U)$: $\{(2,-1,1,0),(-1,0,0,0)\}$;
$\boldsymbol{C}(U^{\mathrm{T}})$: $\{(1,2,0,1),(0,1,1,0)\}$; $\boldsymbol{N}(U^{\mathrm{T}})$: $\{(0,0,1)\}$.

2.4.5 A times every column of B is zero, so $\boldsymbol{C}(B)$ is contained in the nullspace $\boldsymbol{N}(A)$.

2.4.7 From $Ax = 0$, the row space and the nullspace must be orthogonal. See Chapter 3.

2.4.9 $\begin{bmatrix} 1 & 2 & 4 \end{bmatrix}$; $\begin{bmatrix} 1 & 2 & 4 \\ 2 & 4 & 8 \\ 3 & 6 & 12 \end{bmatrix}$ has the same nullspace.

2.4.11 If $Ax = 0$ has a nonzero solution, then $r < n$ and $\boldsymbol{C}(A^{\mathrm{T}})$ is smaller than $\mathbf{R}^n$. So $A^{\mathrm{T}}y = f$ is not solvable for some f. Example: $A = \begin{bmatrix} 1 & 1 \end{bmatrix}$ and $f = (1,2)$.

2.4.13 $d = bc/a$; the only pivot is a.

2.4.15 With independent columns: rank n; nullspace $= \{0\}$; row space is $\mathbf{R}^n$; there exists a *left*-inverse.

2.4.17 $A = \begin{bmatrix} 1 & 1 & 0 \end{bmatrix}$; $B = \begin{bmatrix} 0 & 0 & 1 \end{bmatrix}$.

2.4.19 No. For example, all invertible n by n matrices have the same four subspaces.

2.4.21 (a) $\begin{bmatrix} 1 & 0 \\ 1 & 0 \\ 0 & 1 \end{bmatrix}$ (b) Impossible: dimensions $1 + 1 \neq 3$

(c) $\begin{bmatrix} 1 & 1 \end{bmatrix}$ (d) $\begin{bmatrix} -9 & -3 \\ 3 & 1 \end{bmatrix}$

(e) Impossible: If the row space is equal to the column space, then $m = n$, so $m - r = n - r$.

2.4.23 If A is invertible, then a row space basis and a column space basis is $\{(1,0,0),(0,1,0),(0,0,1)\}$; nullspace and left nullspace bases are empty.
For B, a row space basis is $\{(1,0,0,1,0,0),(0,1,0,0,1,0),(0,0,1,0,0,1)\}$;
a column space basis is $\{(1,0,0),(0,1,0),(0,0,1)\}$; a nullspace basis is $\{(-1,0,0,1,0,0),(0,-1,0,0,1,0),(0,0,-1,0,0,1)\}$; the left nullspace basis is empty.

2.4.25 (a) The two matrices have the same row space and nullspace. Therefore the rank (dimension of the row space) is the same

(b) The matrices have the same column space and left nullspace, so they have the same rank (dimension of the column space).

2.4.27 (a) If $Ax = b$ has no solution, then $r < m$. Of course $r \leq n$. We cannot compare m and n.

(b) If $m - r > 0$, the nullspace of A^{T} contains a nonzero vector.

2.4.29 Row space basis $\{(1,2,3,4),(0,1,2,3),(0,0,1,2)\}$; nullspace basis $\{(0,1,-2,1)\}$; column space basis $\{(1,0,0),(0,1,0),(0,0,1)\}$; left nullspace has empty basis.

2.4.31 If $Av = 0$ and v is a row of A then $v \cdot v = 0$. Only $v = 0$ is in both spaces.

2.4.33 Row $3 - 2\,(\text{row } 2) + \text{row } 1 = \text{zero row}$, so the vectors $c\,(1, -2, 1)$ are in the left nullspace. The same vectors happen to be in the nullspace.

2.4.35 (a) u and w span $\boldsymbol{C}(A)$.

(b) v and z span $\boldsymbol{C}(A^{\mathrm{T}})$.

(c) $\operatorname{rank}(A) < 2$ if u and w are dependent or v and z are dependent.

(d) The rank of $uv^{\mathrm{T}} + wz^{\mathrm{T}}$ is 2.

2.4.37 (a) True (same rank).

(b) False: $A = \begin{bmatrix} 1 & 0 \end{bmatrix}$.

(c) False (A can be invertible and also unsymmetric).

(d) True.

2.4.39 Yes. One possibility is $\begin{bmatrix} 1 & & \\ & & \\ & & \end{bmatrix}, \begin{bmatrix} 1 & 0 & \\ & & \\ & & \end{bmatrix}, \begin{bmatrix} 1 & 0 & 1 \\ & & \\ & & \end{bmatrix}, \begin{bmatrix} 1 & 0 & 1 \\ & 0 & \\ & & \end{bmatrix}, \begin{bmatrix} 1 & 0 & 1 \\ & 0 & \\ & 1 & \end{bmatrix}, \begin{bmatrix} 1 & 0 & 1 \\ & 0 & \\ 0 & 1 & \end{bmatrix},$

$\begin{bmatrix} 1 & 0 & 1 \\ & 0 & 1 \\ 0 & 1 & \end{bmatrix}, \begin{bmatrix} 1 & 0 & 1 \\ & 0 & 1 \\ 0 & 1 & 0 \end{bmatrix}, \begin{bmatrix} 1 & 0 & 1 \\ 1 & 0 & 1 \\ 0 & 1 & 0 \end{bmatrix}.$

2.4.41 If the rank is $r = n$, then the nullspace is the zero vector and $x_n = 0$.

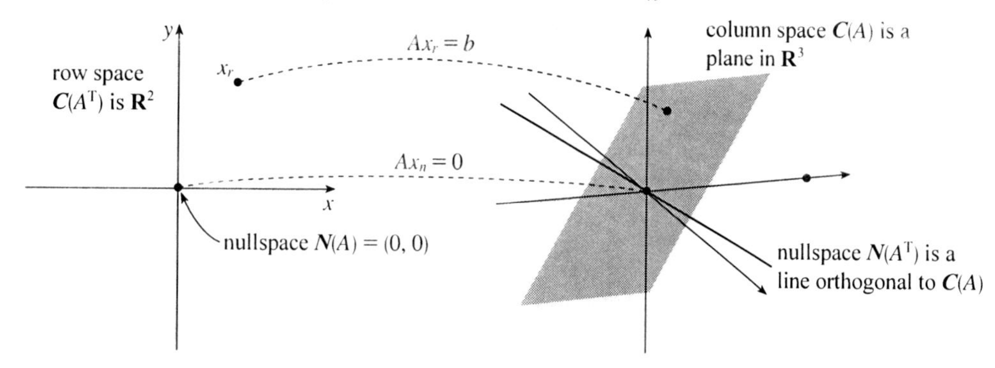

Problem Set 2.5, page 122

2.5.1 $A = \begin{bmatrix} 1 & -1 & 0 \\ 0 & 1 & -1 \\ 1 & 0 & -1 \end{bmatrix}$; $\boldsymbol{N}(A)$ contains multiples of $\begin{bmatrix} 1 \\ 1 \\ 1 \end{bmatrix}$; $\boldsymbol{N}(A^{\mathrm{T}})$ contains multiples of $\begin{bmatrix} 1 \\ 1 \\ -1 \end{bmatrix}$.

2.5.3 The entries in each row add to zero. Therefore any combination will have that same property: $f_1 + f_2 + f_3 = 0$; $A^{\mathrm{T}}y = f \Rightarrow y_1 + y_3 = f_1, -y_1 + y_2 = f_2, -y_2 - y_3 = f_3 \Rightarrow f_1 + f_2 + f_3 = 0$. It means that the total current entering from outside is zero.

2.5.5 $\begin{bmatrix} c_1 + c_3 & -c_1 & -c_3 \\ -c_1 & c_1 + c_2 & -c_2 \\ -c_3 & -c_2 & c_2 + c_3 \end{bmatrix}$; $\begin{bmatrix} c_1 + c_3 & -c_1 \\ -c_1 & c_1 + c_2 \end{bmatrix}$ has pivots $c_1 + c_3$, $\dfrac{c_1c_2 + c_1c_3 + c_2c_3}{c_1 + c_3}$.

2.5.7 Conditions on b are $b_1 + b_4 - b_5 = 0$, $b_3 - b_4 + b_6 = 0$, $b_2 - b_5 + b_6 = 0$.

2.5.9 $\begin{bmatrix} 3 & -1 & -1 & -1 \\ -1 & 3 & -1 & -1 \\ -1 & -1 & 3 & -1 \\ -1 & -1 & -1 & 3 \end{bmatrix}$, $\begin{bmatrix} c_1+c_2+c_5 & -c_1 & -c_2 & -c_5 \\ -c_1 & c_1+c_3+c_4 & -c_3 & -c_4 \\ -c_2 & -c_3 & c_2+c_3+c_6 & -c_6 \\ -c_5 & -c_4 & -c_6 & c_4+c_5+c_6 \end{bmatrix}$

Those c's that connect to node j will appear in row j.

2.5.11 $$\begin{bmatrix} 1 & 0 & 0 & 0 & -1 & 1 & 0 \\ 0 & \frac{1}{2} & 0 & 0 & -1 & 0 & 1 \\ 0 & 0 & \frac{1}{2} & 0 & 0 & 1 & 0 \\ 0 & 0 & 0 & 1 & 0 & 0 & -1 \\ -1 & -1 & 0 & 0 & 0 & 0 & 0 \\ 1 & 0 & 1 & 0 & 0 & 0 & 0 \\ 0 & 1 & 0 & -1 & 0 & 0 & 0 \end{bmatrix} \begin{bmatrix} y_1 \\ y_2 \\ y_3 \\ y_4 \\ x_1 \\ x_2 \\ x_3 \end{bmatrix} = \begin{bmatrix} 0 \\ 0 \\ 0 \\ 0 \\ f_1 \\ f_2 \\ f_3 \end{bmatrix}; x = \begin{bmatrix} -4 \\ -\frac{5}{3} \\ -\frac{14}{3} \\ 0 \end{bmatrix}; y = \begin{bmatrix} -\frac{7}{3} \\ \frac{4}{3} \\ \frac{10}{3} \\ \frac{14}{3} \end{bmatrix}.$$

2.5.13 There are 20 choices of 3 edges out of 6 because $\binom{6}{3} = \frac{6!}{3!\,3!} = 20$. Four choices give triangles, leaving 16 spanning trees.

2.5.15 The strength of the opposition is already built in (for connected graphs).

2.5.17 9 nodes $-$ 12 edges $+$ 4 loops $= 1$; 7 nodes $-$ 12 edges $+$ 6 loops $= 1$

2.5.19 $x = (1, 1, 1, 1)$ gives $Ax = 0$ so then $A^{\mathrm{T}}Ax = 0$; rank is again $n - 1$.

2.5.21 $M = \begin{bmatrix} 0 & 1 & 1 & 1 \\ 1 & 0 & 1 & 1 \\ 1 & 1 & 0 & 1 \\ 1 & 1 & 1 & 0 \end{bmatrix}$ and $M^2 = \begin{bmatrix} 3 & 2 & 2 & 2 \\ 2 & 3 & 2 & 2 \\ 2 & 2 & 3 & 2 \\ 2 & 2 & 2 & 3 \end{bmatrix}$. $(M^2)_{ij} = a_{i1}a_{1j} + \cdots + a_{in}a_{nj}$ and we get $a_{ik}a_{kj} = 1$ when there is a two-step path $i \to k \to j$. Notice three paths from a node to itself.

Problem Set 2.6, page 133

2.6.1 $\begin{bmatrix} 0 & -1 \\ 1 & 0 \end{bmatrix}$ rotates through 90° and $\begin{bmatrix} 1 & 0 \\ 0 & 0 \end{bmatrix}$ projects onto the x-axis, so

$\begin{bmatrix} 1 & 0 \\ 0 & 0 \end{bmatrix}\begin{bmatrix} 0 & -1 \\ 1 & 0 \end{bmatrix} = \begin{bmatrix} 0 & -1 \\ 0 & 0 \end{bmatrix}$ does both (in that order).

Projection onto the x-axis followed by projection onto the y-axis constitutes projection onto the origin: $\begin{bmatrix} 0 & 0 \\ 0 & 1 \end{bmatrix}\begin{bmatrix} 1 & 0 \\ 0 & 0 \end{bmatrix} = \begin{bmatrix} 0 & 0 \\ 0 & 0 \end{bmatrix}$.

2.6.3 $\|Ax\|^2 = 1$ always produces an ellipse.

2.6.5 $(1,0) \to (1,3)$, $(2,0) \to (2,6)$, $(-1,0) \to (-1,-3)$. The x-axis is rotated; vertical lines shift up or down but stay vertical.

2.6.7 The second derivative matrix is $\begin{bmatrix} 0 & 0 & 2 & 0 \\ 0 & 0 & 0 & 6 \\ 0 & 0 & 0 & 0 \\ 0 & 0 & 0 & 0 \end{bmatrix}$. The nullspace is spanned by $(1,0,0,0)$ and $(0,1,0,0)$; it is the subspace of linear polynomials, because second derivatives of linear functions are zero. The column space is accidentally the same as the nullspace, because second derivatives of cubics are linear.

2.6.9 e^t and e^{-t} are a basis for the solutions of $u'' = u$.

2.6.11 $\begin{bmatrix} \cos\theta & -\sin\theta \\ \sin\theta & \cos\theta \end{bmatrix}\begin{bmatrix} \cos\theta & -\sin\theta \\ \sin\theta & \cos\theta \end{bmatrix} = \begin{bmatrix} 1 & 0 \\ 0 & 1 \end{bmatrix}$, so $H^2 = I$.

2.6.13 (a) Yes. (b) Yes. We don't need parentheses $(AB)\,C$ or $A\,(BC)$ for ABC!

2.6.15 $A = \begin{bmatrix} 1 & 0 & 0 & 0 \\ 0 & 0 & 1 & 0 \\ 0 & 1 & 0 & 0 \\ 0 & 0 & 0 & 1 \end{bmatrix}$ and $A^2 = I$; the double transpose of a matrix gives the matrix itself. Note that $A_{23} = 1$ because the transpose of matrix 2 is matrix 3.

2.6.17 $A = \begin{bmatrix} 0 & 0 & 0 \\ 1 & 0 & 0 \\ 0 & 1 & 0 \\ 0 & 0 & 1 \end{bmatrix}$; $B = \begin{bmatrix} 0 & 1 & 0 & 0 \\ 0 & 0 & 1 & 0 \\ 0 & 0 & 0 & 1 \end{bmatrix}$; $AB = \begin{bmatrix} 0 & 0 & 0 & 0 \\ 0 & 1 & 0 & 0 \\ 0 & 0 & 1 & 0 \\ 0 & 0 & 0 & 1 \end{bmatrix}$; $BA = \begin{bmatrix} 1 & 0 & 0 \\ 0 & 1 & 0 \\ 0 & 0 & 1 \end{bmatrix}$.

2.6.19 (a) This transformation is invertible, with $T^{-1}(y) = y^{1/3}$.

(b) $T(x) = e^x = b$ has no solution for $b \le 0$, so T is not invertible.

(c) This transformation is invertible, with $T^{-1}(y) = y - 11$.

(d) This is not invertible; $T(x) = \cos x = b$ has infinitely many solutions if $-1 \le b \le 1$, none otherwise.

2.6.21 With $w = 0$ linearity gives $T(v+0) = T(v) + T(0)$. Thus $T(0) = 0$. With $c = -1$ linearity gives $T(-0) = -T(0)$. Certainly $T(-0) = T(0)$. Thus $T(0) = 0$.

2.6.23 $S(T(v)) = S(v) = v$

2.6.25 (a) This transformation fails $T(2v) = 2T(v)$.

(b) This transformation is linear.

(c) This transformation is linear.

(d) This transformation fails $T(v+w) = T(v) + T(w)$.

2.6.27 $T(T(v)) = (v_3, v_1, v_2)$, $T^3(v) = v$, $T^{100}(v) = T\left(T^{99}(v)\right) = T(v)$.

2.6.29 (a) $T(1,0) = 0$ (b) $(0,0,1)$ is not in the range. (c) $T(0,1) = 0$

2.6.31 The distributive law gives $A(M_1 + M_2) = AM_1 + AM_2$. Commutativity for scalars gives $A(cM) = c(AM)$.

2.6.33 No matrix A gives $A\begin{bmatrix} 0 & 0 \\ 1 & 0 \end{bmatrix} = \begin{bmatrix} 0 & 1 \\ 0 & 0 \end{bmatrix}$.

2.6.35 $T(I) = 0$, but $M = \begin{bmatrix} 0 & b \\ 0 & 0 \end{bmatrix} = T(M)$; these fill the range. $\left\{\begin{bmatrix} a & 0 \\ c & d \end{bmatrix}\right\}$ is the kernel.

2.6.37 (a) $M = \begin{bmatrix} r & s \\ t & u \end{bmatrix}$ (b) $N = \begin{bmatrix} a & b \\ c & d \end{bmatrix}^{-1}$ (c) $ad = bc$

2.6.39 The change-of-basis matrix is a *permutation matrix*. A matrix that changes the lengths of the basis vectors is a *positive diagonal matrix*.

2.6.41 $\begin{bmatrix} 1 & a & a^2 \\ 1 & b & b^2 \\ 1 & c & c^2 \end{bmatrix}\begin{bmatrix} A \\ B \\ C \end{bmatrix} = \begin{bmatrix} 4 \\ 5 \\ 6 \end{bmatrix}$. $\begin{vmatrix} 1 & a & a^2 \\ 1 & b & b^2 \\ 1 & c & c^2 \end{vmatrix} = (b-a)(c-a)(c-b)$; the points a, b, c must be different and then the determinant is nonzero (interpolation is possible).

2.6.43 If T is not invertible then $\{T(v_1), \ldots, T(v_n)\}$ is not a basis. Then we cannot choose $\{w_i\} = \{T(v_i)\}$ as an output basis.

2.6.45 $S(T(v)) = (-1, 2)$ but $S(v) = (-2, 1)$ and $T(S(v)) = (1, -2)$. So $TS \neq ST$.

2.6.47 The Hadamard matrix H has orthogonal columns of length 2, so the inverse of H is $\frac{1}{4}H^{\mathrm{T}} = \frac{1}{4}H$. Thus, $Hx = (7,5,3,1) \Rightarrow x = H^{-1}(7,5,3,1) = (4,1,2,0)$.

2.6.49 False, in general. This is true only if the n nonzero vectors are independent.

Chapter 3 Orthogonality

Problem Set 3.1, page 148

3.1.1 $\|x\| = \sqrt{21}$; $\|y\| = 3\sqrt{2}$; $x^{\mathrm{T}}y = 0$.

3.1.3 $(x_2/x_1)\,(y_2/y_1) = -1$ means that $x_1y_1 + x_2y_2 = 0$, so $x^{\mathrm{T}}y = 0$.

3.1.5 v_1 and v_3 are orthogonal, as are v_2 and v_3.

3.1.7 $x = (-2, 1, 0)$; $y = (-1, -1, 1)$; the row $z = (1, 2, 1)$ is orthogonal to the nullspace.

3.1.9 The orthogonal complement is the line through $(-1, -1, 1)$ and $(0, 0, 0)$.

3.1.11 If $A^{\mathrm{T}}y = 0$ then $y^{\mathrm{T}}b = y^{\mathrm{T}}Ax = \left(y^{\mathrm{T}}A\right)x = 0$, which contradicts $y^{\mathrm{T}}b \neq 0$.

3.1.13 Any $\mathbf{y}$ in $\mathbf{R}^m$ is the sum of a vector in the column space and a vector in the left nullspace.

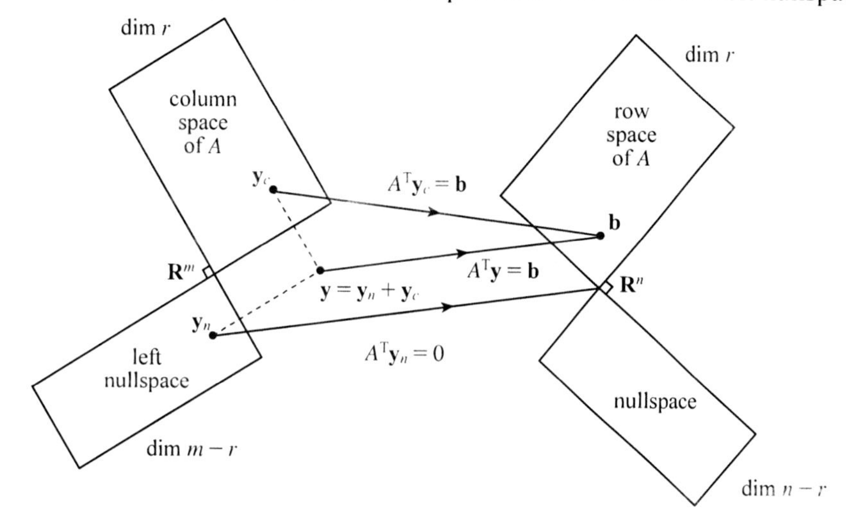

3.1.15 There is no such matrix because $(1, 2, 1)^{\mathrm{T}}(1, -2, 1) \neq 0$.

3.1.17 The matrix with the basis for $\mathbf{V}$ as its rows has nullspace is $\mathbf{V}^{\perp} = \mathbf{W}$.

3.1.19 (a) If $\mathbf{V}$ and $\mathbf{W}$ are lines in $\mathbf{R}^3$, then $\mathbf{V}^{\perp}$ and $\mathbf{W}^{\perp}$ are intersecting planes.

(b) If $\mathbf{V}$ and $\mathbf{Z}$ are the x-axis in $\mathbf{R}^2$ and $\mathbf{W}$ is the y-axis, then $\mathbf{V} \perp \mathbf{W}$ and $\mathbf{W} \perp \mathbf{Z}$ but $\mathbf{V} = \mathbf{Z}$!

3.1.21 $(1, 2, -1)$ is perpendicular to $\mathbf{P}$. $A = \begin{bmatrix} 1 & 1 & 3 \\ 0 & 1 & 2 \end{bmatrix}$ has $\mathbf{N}(A) = \mathbf{P}$; $B = \begin{bmatrix} 1 & 2 & -1 \end{bmatrix}$ has row space $\mathbf{P}$.

3.1.23 The subspaces of $A = \begin{bmatrix} 1 & 1 \\ 2 & 2 \end{bmatrix}$ are four lines. $(1,1)$ is orthogonal to $(-1,1)$ and $(1,2)$ is orthogonal to $(-2,1)$. Of course, the row space is orthogonal to the nullspace.

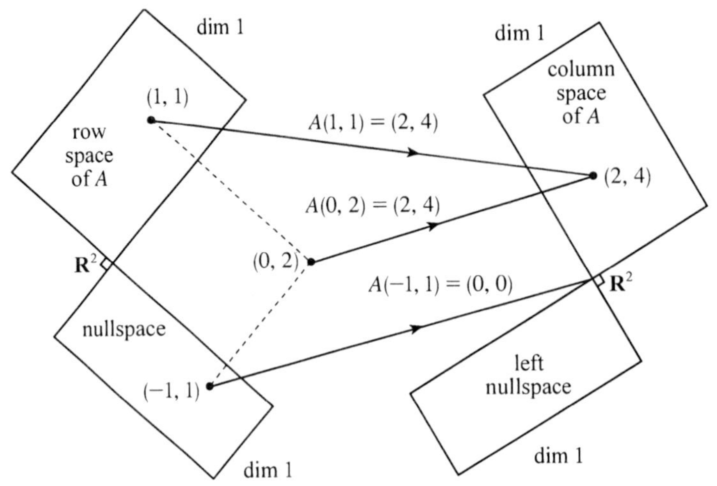

3.1.25 (a) $\begin{bmatrix} 1 & 2 & -3 \\ 2 & -3 & 1 \\ -3 & 5 & -2 \end{bmatrix}$ (b) Impossible; $\begin{bmatrix} 2 \\ -3 \\ 5 \end{bmatrix}$ is not orthogonal to $\begin{bmatrix} 1 \\ 1 \\ 1 \end{bmatrix}$.

(c) It is impossible that $\begin{bmatrix} 1 \\ 1 \\ 1 \end{bmatrix}$ be in $\mathbf{C}(A)$ and $\begin{bmatrix} 1 \\ 0 \\ 0 \end{bmatrix}$ be in $\mathbf{N}\left(A^{\mathrm{T}}\right)$ because these vectors are not perpendicular.

(d) $A = \begin{bmatrix} 1 & -1 \\ 1 & -1 \end{bmatrix}$ has $A^2 = 0$.

(e) If this were possible, then $(1,1,1)$ would be in both the nullspace and the row space. Thus there is no such matrix.

3.1.27 (a) If $Ax = b$ has a solution and $A^{\mathrm{T}}y = 0$, then $b^{\mathrm{T}}y = (Ax)^{\mathrm{T}}y = 0$ and y is perpendicular to b.

(b) If $A^{\mathrm{T}}y = c$ has a solution and $Ax = 0$, then $c^{\mathrm{T}}x = \left(A^{\mathrm{T}}y\right)^{\mathrm{T}}x = 0$ and x is perpendicular to c.

3.1.29 $x = x_r + x_n$, where x_r is in the row space and x_n is in the nullspace. Then $Ax_n = 0$ and $Ax = Ax_r + Ax_n = Ax_r$. All vectors Ax are combinations of the columns of A. If $x = (1,0)$ then $x_r = \left(\frac{1}{2}, \frac{1}{2}\right)$.

3.1.31 (a) For a symmetric matrix the column space and row space are the same.

(b) x is in the nullspace and z is in the column space — which is the same as the row space. So these "eigenvectors" have $x^{\mathrm{T}}z = 0$.

3.1.33 x splits into $x_r + x_n = (1,-1) + (1,1) = (2,0)$.
Two possible illustrations:

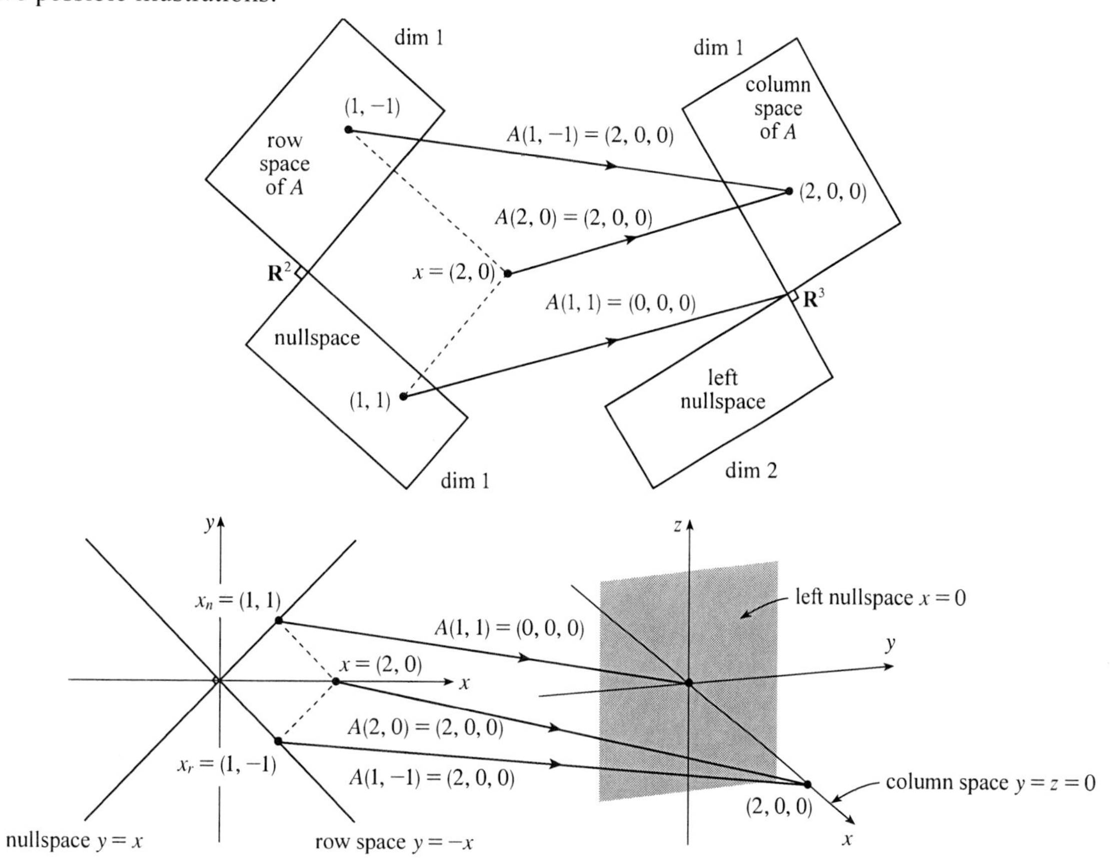

3.1.35 $Ax = B\widehat{x}$ means that $\begin{bmatrix} A & B \end{bmatrix}\left[\begin{smallmatrix} x \\ -\widehat{x} \end{smallmatrix}\right] = 0$. Three homogeneous equations in four unknowns always have a nonzero solution. Here $x = (3,1)$ and $\widehat{x} = (1,0)$, and $Ax = B\widehat{x} = (5,6,5)$ is in both column spaces. Two planes through the origin in $\mathbf{R}^3$ must intersect in a line, at least!

3.1.37 $A^{\mathrm{T}}y = 0$ gives $(Ax)^{\mathrm{T}}y = x^{\mathrm{T}}A^{\mathrm{T}}y = 0$. Then $y \perp Ax$ and $\mathbf{N}\left(A^{\mathrm{T}}\right) \perp \mathbf{C}(A)$.

3.1.39 $\mathbf{S}^{\perp}$ is the nullspace of $A = \begin{bmatrix} 1 & 5 & 1 \\ 2 & 2 & 2 \end{bmatrix}$. Therefore $\mathbf{S}^{\perp}$ is a subspace even if $\mathbf{S}$ is not.

3.1.41 If $\mathbf{V}$ is all of $\mathbf{R}^4$, then $\mathbf{V}^{\perp}$ contains only the *zero vector*. Then $\left(\mathbf{V}^{\perp}\right)^{\perp} = \mathbf{R}^4 = \mathbf{V}$.

3.1.43 $(1,1,1,1)$ is a basis for $\mathbf{P}^{\perp}$. $A = \begin{bmatrix} 1 & 1 & 1 & 1 \end{bmatrix}$ has the plane $\mathbf{P}$ as its nullspace.

3.1.45 Column 1 of A^{-1} is orthogonal to the space spanned by the 2nd, ..., nth rows of A.

3.1.47 $A = \begin{bmatrix} 2 & 2 & -1 \\ -1 & 2 & 2 \\ 2 & -1 & 2 \end{bmatrix}$, $A^{\mathrm{T}}A = 9I$ is diagonal because $\left(A^{\mathrm{T}}A\right)_{ij} = (\text{column } i \text{ of } A) \cdot (\text{column } j)$.

3.1.49 (a) $(1,-1,0)$ is in both planes. Normal vectors are perpendicular, perpendicular planes still intersect!

(b) We need *three* orthogonal vectors to span the whole orthogonal complement in $\mathbf{R}^5$.

(c) Lines can meet without being orthogonal.

3.1.51 When $AB = 0$, the column space of B is contained in the nullspace of A. Therefore the $\dim(\mathbf{C}(B)) \le \dim(\mathbf{N}(A))$. This means that $\operatorname{rank}(B) \le 4 - \operatorname{rank}(A)$.

Problem Set 3.2, page 157

3.2.1 (a) $\frac{1}{2}(x+y) \geq \sqrt{xy}$. That is, the arithmetic mean of x and y is larger than or equal to their geometric mean.

(b) $\|x+y\|^2 \leq (\|x\|+\|y\|)^2$ means that $(x+y)^{\mathrm{T}}(x+y) \leq \|x\|^2 + 2\|x\|\,\|y\| + \|y\|^2$. The left side is $x^{\mathrm{T}}x + 2x^{\mathrm{T}}y + y^{\mathrm{T}}y$. After cancelling, this is $x^{\mathrm{T}}y \leq \|x\|\,\|y\|$.

3.2.3 $p = \left(\frac{10}{3}, \frac{10}{3}, \frac{10}{3}\right); \left(\frac{5}{9}, \frac{10}{9}, \frac{10}{9}\right)$.

3.2.5 $\cos\theta = \dfrac{1}{\sqrt{n}}$, so $\theta = \arccos\dfrac{1}{\sqrt{n}}$; $P = \begin{bmatrix} 1 \\ \vdots \\ 1 \end{bmatrix}\begin{bmatrix} \frac{1}{n} & \cdots & \frac{1}{n} \end{bmatrix} = \begin{bmatrix} \frac{1}{n} & \cdots & \frac{1}{n} \\ \vdots & \ddots & \vdots \\ \frac{1}{n} & \cdots & \frac{1}{n} \end{bmatrix}$.

3.2.7 Choose $b = (1, \ldots, 1)$. There is equality if $a_1 = \cdots = a_n$, in which case a is parallel to b.

3.2.9 $P^2 = \dfrac{aa^{\mathrm{T}}aa^{\mathrm{T}}}{a^{\mathrm{T}}aa^{\mathrm{T}}a} = \dfrac{a\left(a^{\mathrm{T}}a\right)a^{\mathrm{T}}}{\left(a^{\mathrm{T}}a\right)\left(a^{\mathrm{T}}a\right)} = \dfrac{aa^{\mathrm{T}}}{a^{\mathrm{T}}a} = P.$

3.2.11 (a) $P = \begin{bmatrix} \frac{1}{10} & \frac{3}{10} \\ \frac{3}{10} & \frac{9}{10} \end{bmatrix}$; $P_2 = I - P_1 = \begin{bmatrix} \frac{9}{10} & -\frac{3}{10} \\ -\frac{3}{10} & \frac{1}{10} \end{bmatrix}$

(b) $P_1 + P_2 = \begin{bmatrix} 1 & 0 \\ 0 & 1 \end{bmatrix}$; $P_1P_2 = \begin{bmatrix} 0 & 0 \\ 0 & 0 \end{bmatrix}$. The sum of the projections onto two perpendicular lines gives the vector itself. The projection onto one line and then a perpendicular line gives the zero vector.

3.2.13 Trace $= \dfrac{a_1a_1}{a^{\mathrm{T}}a} + \cdots + \dfrac{a_na_n}{a^{\mathrm{T}}a} = \dfrac{a^{\mathrm{T}}a}{a^{\mathrm{T}}a} = 1.$

3.2.15 $\|Ax\|^2 = (Ax)^{\mathrm{T}}(Ax) = xA^{\mathrm{T}}Ax$, $\left\|A^{\mathrm{T}}x\right\|^2 = \left(A^{\mathrm{T}}x\right)^{\mathrm{T}}\left(A^{\mathrm{T}}x\right) = xAA^{\mathrm{T}}x$. If $A^{\mathrm{T}}A = AA^{\mathrm{T}}$ then $\|Ax\| = \left\|A^{\mathrm{T}}x\right\|$. (These matrices are called *normal*.)

3.2.17 (a) $\dfrac{a^{\mathrm{T}}b}{a^{\mathrm{T}}a} = \dfrac{5}{3}$; $p = \left(\frac{5}{3}, \frac{5}{3}, \frac{5}{3}\right)$; $e = \left(-\frac{2}{3}, \frac{1}{3}, \frac{1}{3}\right)$ has $e^{\mathrm{T}}a = 0$.

(b) $\dfrac{a^{\mathrm{T}}b}{a^{\mathrm{T}}a} = -1$; $p = (1, 3, 1) = b$ and $e = (0, 0, 0)$.

3.2.19 $P_1 = \dfrac{1}{3}\begin{bmatrix} 1 & 1 & 1 \\ 1 & 1 & 1 \\ 1 & 1 & 1 \end{bmatrix} = P_1^2$ and $P_1b = \dfrac{1}{3}\begin{bmatrix} 5 \\ 5 \\ 5 \end{bmatrix}$. $P_2 = \dfrac{1}{11}\begin{bmatrix} 1 & 3 & 1 \\ 3 & 9 & 3 \\ 1 & 3 & 1 \end{bmatrix}$ and $P_2b = \begin{bmatrix} 1 \\ 3 \\ 1 \end{bmatrix}$.

3.2.21 $P_1 = \dfrac{1}{9}\begin{bmatrix} 1 & -2 & -2 \\ -2 & 4 & 4 \\ -2 & 4 & 4 \end{bmatrix}$, $P_2 = \dfrac{1}{9}\begin{bmatrix} 4 & 4 & -2 \\ 4 & 4 & -2 \\ -2 & -2 & 1 \end{bmatrix}$. $P_1P_2 = 0$ because $a_1 \perp a_2$.

3.2.23 $P_1 + P_2 + P_3 = \dfrac{1}{9}\begin{bmatrix} 1 & -2 & -2 \\ -2 & 4 & 4 \\ -2 & 4 & 4 \end{bmatrix} + \dfrac{1}{9}\begin{bmatrix} 4 & 4 & -2 \\ 4 & 4 & -2 \\ -2 & -2 & 1 \end{bmatrix} + \dfrac{1}{9}\begin{bmatrix} 4 & -2 & 4 \\ -2 & 1 & -2 \\ 4 & -2 & 4 \end{bmatrix} = I.$

3.2.25 Since A is invertible, $P = A\left(A^{\mathrm{T}}A\right)^{-1}A^{\mathrm{T}} = AA^{-1}\left(A^{\mathrm{T}}\right)^{-1}A^{\mathrm{T}} = I$. It is a projection onto all of $\mathbf{R}^2$.

Problem Set 3.3, page 170

3.3.1 $\widehat{x} = 2$; $E^2 = (10 - 3x)^2 + (5 - 4x)^2$ is minimized; $(4, -3)^{\mathrm{T}}(3, 4) = 0$.

3.3.3 $\widehat{x} = \begin{bmatrix} \frac{1}{3} \\ \frac{1}{3} \end{bmatrix}$; $p = \begin{bmatrix} \frac{1}{3} \\ \frac{1}{3} \\ \frac{2}{3} \end{bmatrix}$; $b - p = \begin{bmatrix} \frac{2}{3} \\ \frac{2}{3} \\ -\frac{2}{3} \end{bmatrix}$ is perpendicular to both columns.

3.3.5 $b = 4, 5, 9$ at $t = -1, 0, 1$; the best line is $y = 6 + \frac{5}{2}t$; $p = \left(\frac{7}{2}, 6, \frac{17}{2}\right)$.

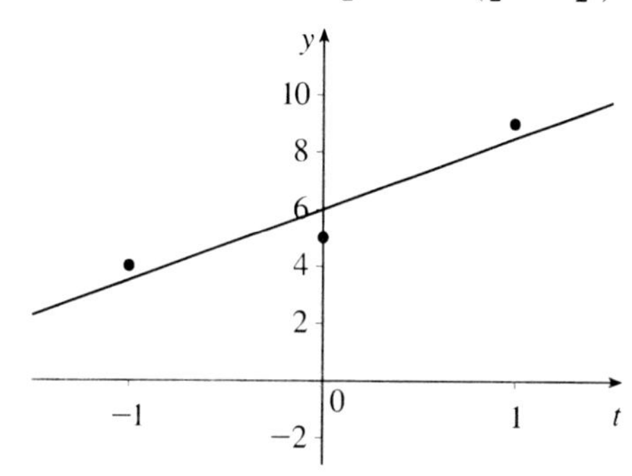

3.3.7 $P = A\left(A^{\mathrm{T}}A\right)^{-1}A^{\mathrm{T}} = \begin{bmatrix} 1 & \frac{1}{2} & 0 \\ \frac{1}{2} & \frac{1}{2} & -\frac{1}{2} \\ 0 & -\frac{1}{2} & 1 \end{bmatrix}$

3.3.9 (a) $P^{\mathrm{T}} = \left(P^{\mathrm{T}}P\right)^{\mathrm{T}} = P$. Then $P = P^{\mathrm{T}}P = P^2$. (b) P projects onto the space $\mathbf{Z} = \{0\}$.

3.3.11 $P + Q = I$, $PQ = 0$, transpose to $QP = 0$, so $(P - Q)(P - Q) = P - 0 - 0 + Q = I$.

3.3.13 Best line $y = \frac{61}{35} - \frac{36}{35}t$; $p = \left(\frac{133}{35}, \frac{95}{35}, \frac{61}{35}, -\frac{11}{35}\right) = \left(\frac{19}{5}, \frac{19}{7}, \frac{61}{35}, -\frac{11}{35}\right)$ from $C + Dt$.

3.3.15 $H^2 = (I - 2P)^2 = I - 4P + 4P^2 = I - 4P + 4P = I$. Two reflections give I.

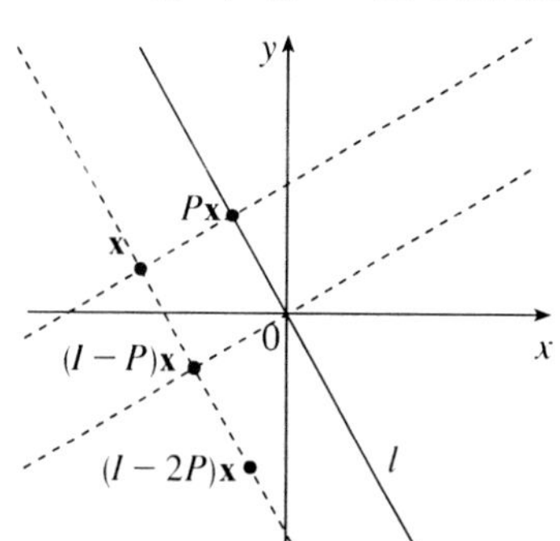

3.3.17 The projection onto $x + y = 0$ is equal to the projection onto $(-1, 1)$, namely $\begin{bmatrix} \frac{1}{2} & -\frac{1}{2} \\ -\frac{1}{2} & \frac{1}{2} \end{bmatrix}$.

3.3.19 If the rows are independent, the projection matrix onto the row space is be $A^{\mathrm{T}}\left(AA^{\mathrm{T}}\right)^{-1}A$.

3.3.21 The best line is $\begin{bmatrix} a_1^{\mathrm{T}}a_1 & -a_1^{\mathrm{T}}a_2 \\ -a_1^{\mathrm{T}}a_2 & a_2^{\mathrm{T}}a_2 \end{bmatrix}\begin{bmatrix} \widehat{x}_1 \\ \widehat{x}_2 \end{bmatrix} = \begin{bmatrix} a_1^{\mathrm{T}}b \\ -a_2^{\mathrm{T}}b \end{bmatrix}$. If $a_1 = (1, 1, 0)$, $a_2 = (0, 1, 0)$, and $(b = 2, 1, 4)$, then $\widehat{x} = \begin{bmatrix} 2 \\ 1 \end{bmatrix}$.

3.3.23 If $C = (A^{\mathrm{T}}A)^{-1}A^{\mathrm{T}}b$, $A^{\mathrm{T}} = \begin{bmatrix} 1 & \cdots & 1 \end{bmatrix}$, and $b = (y_1, \cdots, y_m)^{\mathrm{T}}$, then $C = \dfrac{A^{\mathrm{T}}b}{A^{\mathrm{T}}A} = \dfrac{y_1 + \cdots + y_m}{m}$.

3.3.25 $A = \begin{bmatrix} 1 & -1 & 1 \\ 1 & 0 & 0 \\ 1 & 1 & 1 \\ 1 & 2 & 4 \end{bmatrix}$; $x = \begin{bmatrix} C \\ D \\ E \end{bmatrix}$; $b = \begin{bmatrix} 2 \\ 0 \\ -3 \\ -5 \end{bmatrix}$.

3.3.27 (a) $a^{\mathrm{T}}a = m$, $a^{\mathrm{T}}b = b_1 + \cdots + b_m$. Therefore $\widehat{x}$ is the mean of the b's.

(b) The variance is $\|e\|^2 = \sum_{i=1}^{m}(b_i - \widehat{x})^2$.

(c) $p = (3,3,3)$, $e = (-2,-1,3)$, and $p^{\mathrm{T}}e = 0$. $P = \dfrac{1}{3}\begin{bmatrix} 1 & 1 & 1 \\ 1 & 1 & 1 \\ 1 & 1 & 1 \end{bmatrix}$.

3.3.29 $(\widehat{x} - x)(\widehat{x} - x)^{\mathrm{T}} = (A^{\mathrm{T}}A)^{-1}A^{\mathrm{T}}\left[(b - Ax)(b - Ax)^{\mathrm{T}}\right]A(A^{\mathrm{T}}A)^{-1}$. For independent errors, substituting $(b - Ax)(b - Ax)^{\mathrm{T}} = \sigma^2 I$ gives the *covariance matrix* $(A^{\mathrm{T}}A)^{-1}A^{\mathrm{T}}\sigma^2 A(A^{\mathrm{T}}A)^{-1}$. This simplifies to the neat formula $\sigma^2(A^{\mathrm{T}}A)^{-1}$.

3.3.31 $\frac{1}{10}b_{10} + \frac{9}{10}\widehat{x}_9 = \frac{1}{10}(b_1 + \cdots + b_{10})$

3.3.33 $\begin{bmatrix} 1 & 0 \\ 1 & 1 \\ 1 & 3 \\ 1 & 4 \end{bmatrix}\begin{bmatrix} C \\ D \end{bmatrix} = \begin{bmatrix} 0 \\ 8 \\ 8 \\ 20 \end{bmatrix}$. Change the right side to $p = \begin{bmatrix} 1 \\ 5 \\ 13 \\ 17 \end{bmatrix}$; $\widehat{x} = \begin{bmatrix} 1 \\ 4 \end{bmatrix}$ solves $A\widehat{x} = b$.

3.3.35 Closest parabola: $\begin{bmatrix} 1 & 0 & 0 \\ 1 & 1 & 1 \\ 1 & 3 & 9 \\ 1 & 4 & 16 \end{bmatrix}\begin{bmatrix} C \\ D \\ E \end{bmatrix} = \begin{bmatrix} 0 \\ 8 \\ 8 \\ 20 \end{bmatrix}$. $A^{\mathrm{T}}A\widehat{x} = \begin{bmatrix} 4 & 8 & 26 \\ 8 & 26 & 92 \\ 26 & 92 & 338 \end{bmatrix}\begin{bmatrix} \widehat{C} \\ \widehat{D} \\ \widehat{E} \end{bmatrix} = \begin{bmatrix} 36 \\ 112 \\ 400 \end{bmatrix}$.

3.3.37 (a) The best line is $x = 1 + 4t$, which goes through the center point $(\widehat{t}, \widehat{b}) = (2, 9)$.

(b) From the first equation, $Cm + D\sum t_i = \sum b_i$. Divide by m to get $C + D\widehat{t} = \widehat{b}$.

3.3.39 $\widehat{x}_W = \dfrac{w_1^2 b_1 + \cdots + w_m^2 b_m}{w_1^2 + \cdots + w_m^2}$.

3.3.41 $\widehat{x}_W = \left(\frac{1}{21}, \frac{4}{7}\right)$; $A\widehat{x}_W = \left(\frac{1}{21}, \frac{13}{21}, \frac{25}{21}\right)$, $b - A\widehat{x}_W = \left(-\frac{1}{21}, \frac{8}{21}, -\frac{4}{21}\right)$, $(A\widehat{x}_W)W^{\mathrm{T}}W(b - A\widehat{x}_W) = 0$.

Problem Set 3.4, page 185

3.4.1 (a) $-4 = C - 2D$, $-3 = C - D$, $-1 = C + D$, $0 = C + 2D$.

(b) The best line $y = -2 + t$ goes through all four points; $E^2 = 0$.

(c) b is in the column space.

3.4.3 The projection on a_3 is $\left(-\frac{2}{3}, \frac{1}{3}, -\frac{2}{3}\right)$; the sum is b itself. Notice that $a_1a_1^{\mathrm{T}}$, $a_2a_2^{\mathrm{T}}$, and $a_3a_3^{\mathrm{T}}$ are projections onto three orthogonal directions. Their sum corresponds to projection onto the whole space — that is, identity.

3.4.5 $(I-2uu^{\mathrm{T}})^{\mathrm{T}}(I-2uu^{\mathrm{T}}) = I - 4uu^{\mathrm{T}} + 4uu^{\mathrm{T}}uu^{\mathrm{T}} = I$; $Q = \begin{bmatrix} \frac{1}{2} & -\frac{1}{2} & \frac{1}{2} & \frac{1}{2} \\ -\frac{1}{2} & \frac{1}{2} & \frac{1}{2} & \frac{1}{2} \\ \frac{1}{2} & \frac{1}{2} & \frac{1}{2} & -\frac{1}{2} \\ \frac{1}{2} & \frac{1}{2} & -\frac{1}{2} & \frac{1}{2} \end{bmatrix}$.

3.4.7 $(x_1q_1 + \cdots + x_nq_n)^{\mathrm{T}}(x_1q_1 + \cdots + x_nq_n) = x_1^2 + \cdots + x_n^2 \Rightarrow \|b\|^2 = b^{\mathrm{T}}b = x_1^2 + \cdots + x_n^2$.

3.4.9 The combination closest to q_3 is $0q_1 + 0q_2$.

3.4.11 Q is upper triangular: column 1 has $q_{11} = \pm 1$; by orthogonality column 2 must be $(0, \pm 1, 0, \ldots)$. By orthogonality, column 3 is $(0, 0, \pm 1, 0, \ldots)$, and so on.

3.4.13 $A = \begin{bmatrix} 0 & 0 & 1 \\ 0 & 1 & 1 \\ 1 & 1 & 1 \end{bmatrix} = \begin{bmatrix} 0 & 0 & 1 \\ 0 & 1 & 0 \\ 1 & 0 & 0 \end{bmatrix} \begin{bmatrix} 1 & 1 & 1 \\ 0 & 1 & 1 \\ 0 & 0 & 1 \end{bmatrix} = QR.$

3.4.15 $q_1 = \begin{bmatrix} \frac{1}{3} \\ \frac{2}{3} \\ -\frac{2}{3} \end{bmatrix}$, $q_2 = \begin{bmatrix} \frac{2}{3} \\ \frac{1}{3} \\ \frac{2}{3} \end{bmatrix}$, $q_3 = \begin{bmatrix} -\frac{2}{3} \\ \frac{2}{3} \\ \frac{1}{3} \end{bmatrix}$ is in the left nullspace; $\widehat{x} = \begin{bmatrix} q_1^{\mathrm{T}}b \\ q_2^{\mathrm{T}}b \end{bmatrix} = \begin{bmatrix} 1 \\ 2 \end{bmatrix}$.

3.4.17 $R\widehat{x} = Q^{\mathrm{T}}b$ gives $\begin{bmatrix} 3 & 3 \\ 0 & \sqrt{2} \end{bmatrix} \widehat{x} = \begin{bmatrix} \frac{5}{3} \\ 0 \end{bmatrix}$, and $\widehat{x} = \begin{bmatrix} \frac{5}{9} \\ 0 \end{bmatrix}$.

3.4.19 $C^* - (q_2^{\mathrm{T}}C^*)\,q_2$ is $c - (q_1^{\mathrm{T}}c)\,q_1 - (q_2^{\mathrm{T}}c)\,q_2$ because $q_2^{\mathrm{T}}q_1 = 0$.

3.4.21 By orthogonality, the closest functions are $0 \sin 2x = 0$ and $0 + 0x = 0$.

3.4.23 $a_0 = \frac{1}{2}$, $a_1 = 0$, $b_1 = \frac{2}{\pi}$.

3.4.25 The closest line is $y = \frac{1}{3}$. The line is horizontal because $(x, x^2) = 0$.

3.4.27 $\left(\frac{1}{\sqrt{2}}, -\frac{1}{\sqrt{2}}, 0, 0\right)$, $\left(\frac{1}{\sqrt{6}}, \frac{1}{\sqrt{6}}, \frac{2}{\sqrt{6}}, 0\right)$, $\left(-\frac{1}{2\sqrt{3}}, -\frac{1}{2\sqrt{3}}, \frac{1}{2\sqrt{3}}, -\frac{1}{\sqrt{3}}\right)$.

3.4.29 $A = a = (1, -1, 0, 0)$; $B = b - p = \left(\frac{1}{2}, \frac{1}{2}, -1, 0\right)$; $C = c - p_A - p_B = \left(\frac{1}{3}, \frac{1}{3}, \frac{1}{3}, -1\right)$. Notice the pattern in those orthogonal vectors A, B, C. Next $(1,1,1,1)/4$.

3.4.31 (a) True. $Q = Q^{-1} = I$.

(b) True. $Qx = x_1q_1 + x_2q_2$. $\|Qx\|^2 = x_1^2 + x_2^2$ because $q_1^{\mathrm{T}}q_2 = 0$.

Problem Set 3.5, page 196

3.5.1 $F^2 = \begin{bmatrix} 4 & 0 & 0 & 0 \\ 0 & 0 & 0 & 4 \\ 0 & 0 & 4 & 0 \\ 0 & 4 & 0 & 0 \end{bmatrix}$, $F^4 = \begin{bmatrix} 16 & 0 & 0 & 0 \\ 0 & 16 & 0 & 0 \\ 0 & 0 & 16 & 0 \\ 0 & 0 & 0 & 16 \end{bmatrix} = 4^2 I.$

3.5.3 The submatrix is F_3.

3.5.5 $e^{ix} = -1$ for $x = (2k+1)\pi$, $e^{i\theta} = i$ for $\theta = 2k\pi + \frac{\pi}{2}$, k an integer.

3.5.7 $c = (1, 0, 1, 0)$

3.5.9 (a) $y = F(1, 0, 0, 0)^{\mathrm{T}} =$ column zero of $F = (1, 1, 1, 1)$ (b) $c = \frac{1}{4}(1, 1, 1, 1)$

3.5.11 $c = \begin{bmatrix}1\\0\\1\\0\end{bmatrix} \to \begin{matrix}c_{even} =\\ c_{odd} =\end{matrix} \begin{bmatrix}1\\1\\0\\0\end{bmatrix} \to \begin{matrix}y' =\\ y'' =\end{matrix} \begin{bmatrix}2\\0\\0\\0\end{bmatrix} \to y = \begin{bmatrix}2\\0\\2\\0\end{bmatrix}.$

3.5.13 $c_0 = \frac{1}{4}(f_0 + f_1 + f_2 + f_3)$, $c_1 = \frac{1}{4}(f_0 - if_1 - f_2 + if_3)$, $c_2 = \frac{1}{4}(f_0 - f_1 + f_2 - f_3)$, $c_3 = \frac{1}{4}(f_0 + if_1 - f_2 - if_3)$; f odd means $f_0 = 0$, $f_2 = 0$, $f_3 = -f_1$. Then $c_0 = 0$, $c_2 = 0$, $c_3 = -c_1$ so c is also odd.

3.5.15 $F^{-1} = \begin{bmatrix}1&&&\\&&1&\\&1&&\\&&&1\end{bmatrix} \frac{1}{2} \begin{bmatrix}1&1&&\\1&i^2&&\\&&1&1\\&&1&i^2\end{bmatrix} \frac{1}{2} \begin{bmatrix}1&&1&\\&1&&1\\1&&-1&\\&-i&&i\end{bmatrix} = \frac{1}{4}F^{\mathrm{H}}.$

3.5.17 $D = \begin{bmatrix}1&&\\&e^{2\pi i/6}&\\&&e^{4\pi i/6}\end{bmatrix}$ and $F_3 = \begin{bmatrix}1&1&1\\1&e^{2\pi i/3}&e^{4\pi i/3}\\1&e^{4\pi i/3}&e^{2\pi i/3}\end{bmatrix}.$

3.5.19 $\Lambda = \begin{bmatrix}1&&&\\&i&&\\&&i^2&\\&&&i^3\end{bmatrix}$; $P = \begin{bmatrix}0&1&0\\0&0&1\\1&0&0\end{bmatrix}$ and P^{T} lead to $\lambda^3 - 1 = 0$.

3.5.21 Eigenvalues $e_0 = 2 - 1 - 1 = 0$, $e_1 = 2 - i - i^3 = 2$, $e_2 = 2 - (-1) - (-1) = 4$, $e_3 = 2 - i^3 - i^9 = 2$. Check trace $0 + 2 + 4 + 2 = 8$.

3.5.23 The four components are $(c_0 + c_2) + (c_1 + c_3)$; then $(c_0 - c_2) + i(c_1 - c_3)$; then $(c_0 + c_2) - (c_1 + c_3)$; then $(c_0 - c_2) - i(c_1 - c_3)$. These steps are the Fast Fourier Transform!

Chapter 4 Determinants

Problem Set 4.2, page 206

4.2.1 $\det(2A) = 8$ and $\det(-A) = (-1)^4 \det A = \frac{1}{2}$ and $\det(A^2) = \frac{1}{4}$ and $\det(A^{-1}) = 2$.

4.2.3 The row operations leave $\det A$ unchanged by Rule 5. Then multiplying a row by -1 (Rule 3) gives the row exchange rule: $\det B = -\det A$.

4.2.5 For the first matrix, two row exchanges will produce the identity matrix. The second matrix needs three row exchanges to reach I.

4.2.7 $\det A = 0$ (singular); $\det U = 16$; $\det U^{\mathrm{T}} = 16$; $\det U^{-1} = \frac{1}{16}$; $\det M = 16$ (two exchanges).

4.2.9 The new determinant is $(1 - ml)(ad - bc)$.

4.2.11 If $|\det Q| \neq 1$, then $\det Q^n = (\det Q)^n$ would blow up or approach zero. But Q^n remains an orthogonal matrix. So $\det Q = \pm 1$.

4.2.13 (a) Rule 3 (factoring -1 from each row) gives $\det(K^{\mathrm{T}}) = (-1)^3 \det K$. Then $-\det K = \det K^{\mathrm{T}} = \det K$ gives $\det K = 0$.

(b) $A = \begin{bmatrix} 0 & 0 & 0 & 1 \\ 0 & 0 & 1 & 0 \\ 0 & -1 & 0 & 0 \\ -1 & 0 & 0 & 0 \end{bmatrix}$ has $\det A = 1$.

4.2.15 Adding every column of A to the first column makes it a zero column, so $\det A = 0$. If every row of A sums to 1, then every row of $A - I$ sums to 0 and $\det(A - I) = 0$.

But $\det A$ need not be 1: $A = \begin{bmatrix} \frac{1}{2} & \frac{1}{2} \\ \frac{1}{2} & \frac{1}{2} \end{bmatrix}$ has $\det(A - I) = 0$, but $\det A = 0 \neq 1$.

4.2.17 $\det(A) = 10$, $\det(A^{-1}) = \frac{1}{10}$, $\det(A - \lambda I) = \lambda^2 - 7\lambda + 10 = 0$ for $\lambda = 5$ and $\lambda = 2$.

4.2.19 Taking determinants gives $(\det C)(\det D) = (-1)^n (\det D)(\det C)$. For even n, the reasoning fails because $(-1)^n = 1$, and the conclusion is wrong.

4.2.21 $$\det(A^{-1}) = \det \begin{bmatrix} \dfrac{d}{ad-bc} & \dfrac{-b}{ad-bc} \\ \dfrac{-c}{ad-bc} & \dfrac{a}{ad-bc} \end{bmatrix} = \frac{ad-bc}{(ad-bc)^2} = \frac{1}{ad-bc}.$$

4.2.23 $$\det \begin{bmatrix} 1 & 2 & 3 & 0 \\ 2 & 6 & 6 & 1 \\ -1 & 0 & 0 & 3 \\ 0 & 2 & 0 & 7 \end{bmatrix} = \det \begin{bmatrix} 1 & 2 & 3 & 0 \\ 0 & 2 & 0 & 1 \\ 0 & 0 & 3 & 2 \\ 0 & 0 & 0 & 6 \end{bmatrix} = 36; \det \begin{bmatrix} 2 & 1 & 1 & 1 \\ 1 & 2 & 1 & 1 \\ 1 & 1 & 2 & 1 \\ 1 & 1 & 1 & 2 \end{bmatrix} = \det \begin{bmatrix} 2 & 1 & 1 & 1 \\ 0 & \frac{3}{2} & \frac{1}{2} & \frac{1}{2} \\ 0 & 0 & \frac{4}{3} & \frac{1}{3} \\ 0 & 0 & 0 & \frac{5}{4} \end{bmatrix} = 5.$$

4.2.25 $\det L = 1$, $\det U = -6$, $\det A = -6$, $\det(U^{-1}L^{-1}) = -\frac{1}{6}$, and $\det(U^{-1}L^{-1}A) = 1$.

4.2.27 row 3 $-$ row 2 $=$ row 2 $-$ row 1, so A is singular.

4.2.29 A is rectangular, so $\det(A^{\mathrm{T}}A) \neq (\det A^{\mathrm{T}})(\det A)$ because the latter quantities are not defined.

4.2.31 The Hilbert determinants are 1, 8×10^{-2}, 4.6×10^{-4}, 1.6×10^{-7}, 3.7×10^{-12}, 5.4×10^{-18}, 4.8×10^{-25}, 2.7×10^{-33}, 9.7×10^{-43}, 2.2×10^{-53}. Pivots are ratios of determinants, so the tenth pivot is near $\dfrac{10^{-53}}{10^{-43}} = 10^{-10}$: very small.

4.2.33 The largest determinants of 0-1 matrices for $n = 1, 2, \ldots$ are $1, 1, 2, 3, 5, 9, 32, 56, 144, 320, \ldots$. See mathworld.wolfram.com/HadamardsMaximumDeterminantProblem.html and also the "On-Line Encyclopedia of Integer Sequences" at www.research.att.com. With -1's and 1's, the largest 4 by 4 determinant (see Hadamard in the index) is 16.

4.2.35 $\det(I + M) = 1 + a + b + c + d$. Subtract row 4 from each of rows 1, 2, and 3. Then subtract $a\,(\text{row } 1) + b\,(\text{row } 2) + c\,(\text{row } 3)$ from row 4. This leaves a triangular matrix with 1, 1, 1, and $1 + a + b + c + d$ on its diagonal.

Problem Set 4.3, page 215

4.3.1 (a) $a_{12}a_{21}a_{34}a_{43} = 1$; *even* so $\det A = 1$. (b) $b_{13}b_{22}b_{31}b_{14} = 18$; *odd* so $\det B = -18$.

4.3.3 (a) True by the product rule (b) False; $\det \begin{bmatrix} 1 & 1 \\ 1 & 1 \end{bmatrix} = 0$ (c) False; $\det \begin{bmatrix} 1 & 1 & 0 \\ 0 & 1 & 1 \\ 1 & 0 & 1 \end{bmatrix} = 2$

4.3.5 The $(1, 1)$ cofactor is F_{n-1}. The $(1, 2)$ cofactor has a 1 in column 1, with cofactor F_{n-2}. Multiply by $(-1)^{1+2}$ and also -1 to find $F_n = F_{n-1} + F_{n-2}$. So the determinants are Fibonacci numbers, offset by 1: F_n is the usual F_{n-1}.

4.3.7 Cofactor expansion: $\det A = 4\,(3) - 4\,(1) + 4\,(-4) - 4\,(1) = -12$.

4.3.9 (a) $(n-1)\,n!$ ($n - 1$ for each term) (b) $\left(1 + \dfrac{1}{2!} + \cdots + \dfrac{1}{(n-1)!}\right) n!$ (c) $\frac{1}{3}\left(n^3 + 2n - 3\right)$

4.3.11 $\begin{bmatrix} 0 & A \\ -B & I \end{bmatrix} \begin{bmatrix} I & 0 \\ B & I \end{bmatrix} = \begin{bmatrix} AB & A \\ 0 & I \end{bmatrix}$, $\det \begin{bmatrix} I & 0 \\ B & I \end{bmatrix} = 1 \Rightarrow \det \begin{bmatrix} 0 & A \\ -B & I \end{bmatrix} = \det \begin{bmatrix} AB & A \\ 0 & I \end{bmatrix} =$ $\det(AB)$. Test: If $A = \begin{bmatrix} 1 & 2 \end{bmatrix}$ and $B = \begin{bmatrix} 1 \\ 2 \end{bmatrix}$, then $\det \begin{bmatrix} 0 & A \\ -B & I \end{bmatrix} = 5 = \det(AB)$. If $A = \begin{bmatrix} 1 \\ 2 \end{bmatrix}$ and $B = \begin{bmatrix} 1 & 2 \end{bmatrix}$, then $\det \begin{bmatrix} 0 & A \\ -B & I \end{bmatrix} = 0 = \det(AB)$ because $\text{rank}(AB) \leq \text{rank}(A) \leq n < m$, so the square matrix is singular.

4.3.13 $\det A = 1 + 18 + 12 - 9 - 4 - 6 = 12$, so rows are independent; $\det B = 0$, so rows are dependent (row 1 + row 2 = row 3); $\det C = -1$, C has independent rows.

4.3.15 Each of the six terms in $\det A$ is zero; the rank is at most 2; column 2 has no pivot.

4.3.17 $a_{11}a_{23}a_{32}a_{44}$ has $-$ and $a_{14}a_{23}a_{32}a_{41}$ has $+$, so $\det A = 0$; $\det B = 2 \cdot 4 \cdot 4 \cdot 2 - 1 \cdot 4 \cdot 4 \cdot 1 = 48$.

4.3.19 (a) If $a_{11} = a_{22} = a_{33} = 0$, then at least four terms are zero. (b) At least fifteen terms are zero.

4.3.21 Some term $a_{1\alpha}a_{2\beta} \cdots a_{n\omega}$ in the big formula is not zero! Move rows 1, 2, ..., n into rows α, β, ..., ω. Then these nonzero a's will be on the main diagonal.

4.3.23 $4!/2 = 12$ even permutations; $\det(I + P_{\text{even}}) = 16$ or 4 or 0 (16 comes from $I + I$).

4.3.25 $C = \begin{bmatrix} 3 & 2 & 1 \\ 2 & 4 & 2 \\ 1 & 2 & 3 \end{bmatrix}$ and $AC^{\mathrm{T}} = \begin{bmatrix} 4 & 0 & 0 \\ 0 & 4 & 0 \\ 0 & 0 & 4 \end{bmatrix} = 4I$. Therefore $A^{-1} = \frac{1}{4}C^{\mathrm{T}}$.

4.3.27 $|B_n| = |A_n| - |A_{n-1}| = (n+1) - n = 1$.

4.3.29 We must choose 1's from columns 2 and 1, columns 4 and 3, and so on. Therefore n must be even to have $\det A_n \neq 0$. The number of exchanges is $\frac{1}{2}n$ so $C_n = (-1)^{n/2}$.

4.3.31 $S_1 = 3$, $S_2 = 8$, $S_3 = 21$. The rule looks like every second number in Fibonacci's sequence 1, 1, 2, **3**, 5, **8**, 13, **21**, 34, **55**, ..., so the guess is $S_4 = 55$. The five nonzero terms in the big formula for S_4 are similar to those in the solution to Problem 39, but with 3's instead of 2's: $81 + 1 - 9 - 9 - 9 = 55$.

4.3.33 Changing 3 to 2 in the corner reduces the determinant F_{2n+2} by $+1$ times the cofactor of that corner entry. This cofactor is the determinant of S_{n-1} (one size smaller) which is F_{2n}. Therefore changing 3 to 2 changes the determinant to $F_{2n+2} - F_{2n}$ which is F_{2n+1}.

4.3.35 (a) Every $\det L = 1$; $\det U_k = \det A_k = 2, 6, -6$ for $k = 1, 2, 3$ (b) Pivots $5, \frac{6}{5}, \frac{7}{6}$

4.3.37 The six terms are correct. Row $1 - 2 \cdot$ row $2 +$ row $3 = 0$, so the matrix is singular.

4.3.39 As in Problem 38, the five nonzero terms are given by (row, column) entries

$$(1,1)\,(2,2)\,(3,3)\,(4,4) + (1,2)\,(2,1)\,(3,4)\,(4,3) - (1,2)\,(2,1)\,(3,3)\,(4,4)$$
$$- (1,1)\,(2,2)\,(3,4)\,(4,3) - (1,1)\,(2,3)\,(3,2)\,(4,4)$$

In this case, because A has 2's and -1's, this sum is

$$(2)\,(2)\,(2)\,(2) + (-1)\,(-1)\,(-1)\,(-1) - (-1)\,(-1)\,(2)\,(2) - (2)\,(2)\,(-1)\,(-1) - (2)\,(-1)\,(-1)\,(2)$$
$$= 16 + 1 - 4 - 4 - 4$$

4.3.41 With $a_{11} = 1$, the $-1, 2, -1$ matrix A has $\det A = 1$ and inverse $\left(A^{-1}\right)_{ij} = n + 1 - \max(i, j)$.

4.3.43 Subtracting 1 from the (n, n) entry subtracts its cofactor C_{nn} from the determinant. That cofactor is $C_{nn} = 1$ (smaller Pascal matrix). Subtracting 1 from 1 leaves 0.

Problem Set 4.4, page 225

4.4.1 $\det A = 20$; $C^{\mathrm{T}} = \begin{bmatrix} 20 & -10 & -12 \\ 0 & 5 & 0 \\ 0 & 0 & 4 \end{bmatrix}$; $AC^{\mathrm{T}} = 20I$; $A^{-1} = \dfrac{1}{20}\begin{bmatrix} 20 & -10 & -12 \\ 0 & 5 & 0 \\ 0 & 0 & 4 \end{bmatrix}$.

4.4.3 $(x, y) = \left(\dfrac{d}{ad - bc}, -\dfrac{c}{ad - bc}\right)$; $(x, y, z) = (3, -1, -2)$.

4.4.5 (a) The area of that parallelogram is $\det \begin{bmatrix} 2 & 2 \\ -1 & 3 \end{bmatrix}$, so the triangle ABC has area $\frac{1}{2}\,4 = 2$.

(b) The triangle $A'B'C'$ has the same area; it is just moved to the origin.

4.4.7 The pivots of A are 2, 3, 6 from determinants 2, 6, 36; the pivots of B are 2, 3, 0.

4.4.9 (a) P^2 takes $(1, 2, 3, 4, 5)$ to $(3, 2, 5, 4, 1)$. (b) P^{-1} takes $(1, 2, 3, 4, 5)$ to $(3, 4, 5, 2, 1)$.

4.4.11 The powers of P are all permutation matrices, so eventually one of those matrices must be repeated. If P^r is the same as P^s, then $P^{r-s} = I$.

4.4.13 (a) $\det A = 3$, $\det B_1 = -6$, $\det B_2 = 3$, so $x_1 = -\frac{6}{3} = -2$ and $x_2 = \frac{3}{3} = 1$.

(b) $|A| = 4$, $|B_1| = 3$, $|B_2| = -2$, $|B_3| = 1$. So $x_1 = \frac{3}{4}$ and $x_2 = -\frac{1}{2}$ and $x_3 = \frac{1}{4}$.

4.4.15 (a) $x_1 = \frac{3}{0}$ and $x_2 = \frac{-2}{0}$: no solution. (b) $x_1 = \frac{0}{0}$ and $x_2 = \frac{0}{0}$: *undetermined.*

4.4.17 If the first column in A is also the right side b then $\det A = \det B_1$. Both B_2 and B_3 are singular since a column is repeated. Therefore $x_1 = \dfrac{B_1}{A} = 1$ and $x_2 = x_3 = 0$.

4.4.19 If all cofactors are 0 (even in a single row or column), then $\det A = 0$ (no inverse). $A = \begin{bmatrix} 1 & 1 \\ 1 & 1 \end{bmatrix}$ has no zero cofactor, but it is not invertible.

4.4.21 If $\det A = 1$ and we know the cofactors, then $C^{\mathrm{T}} = A^{-1}$ and also $\det A^{-1} = 1$. Since A is the inverse of A^{-1}, A must be the cofactor matrix for C.

4.4.23 Knowing C, Problem 22 gives $\det A = (\det C)^{1/(n-1)}$ with $n = 4$. So we can construct $A^{-1} = \dfrac{C^{\mathrm{T}}}{\det A}$ using the known cofactors. Invert to find A.

4.4.25 (a) Cofactors $C_{21} = C_{31} = C_{32} = 0$ (b) $C_{12} = C_{21}$, $C_{31} = C_{13}$, $C_{32} = C_{23}$ make S^{-1} symmetric.

4.4.27 (a) The area is $\left|\begin{smallmatrix} 3 & 2 \\ 1 & 4 \end{smallmatrix}\right| = 10$.

(b) The triangle area is $\frac{1}{2}(10) = 5$.

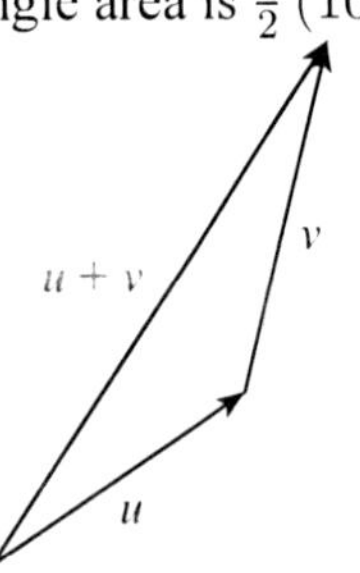

(c) The triangle area is $\frac{1}{2}(10) = 5$.

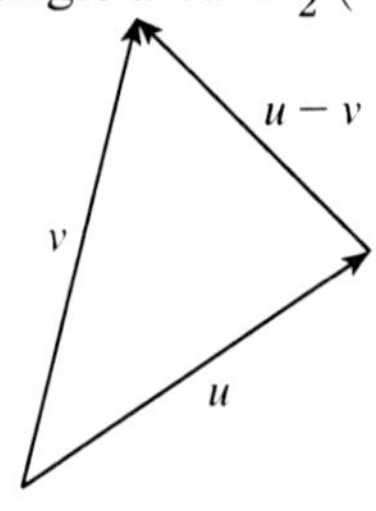

4.4.29 (a) The area is $\dfrac{1}{2}\left|\begin{smallmatrix} 2 & 1 & 1 \\ 3 & 4 & 1 \\ 0 & 5 & 1 \end{smallmatrix}\right| = 5$.

(b) The area is $5 +$ new triangle area $= \dfrac{1}{2}\begin{vmatrix} 2 & 1 & 1 \\ 0 & 5 & 1 \\ -1 & 0 & 1 \end{vmatrix} = 5 + 7 = 12$.

4.4.31 The edges of the hypercube have length $\sqrt{1+1+1+1} = 2$. The volume $\det H$ is $2^4 = 16$. ($H/2$ has orthonormal columns. Then $\det(H/2) = 1$ leads again to $\det H = 16$.)

4.4.33

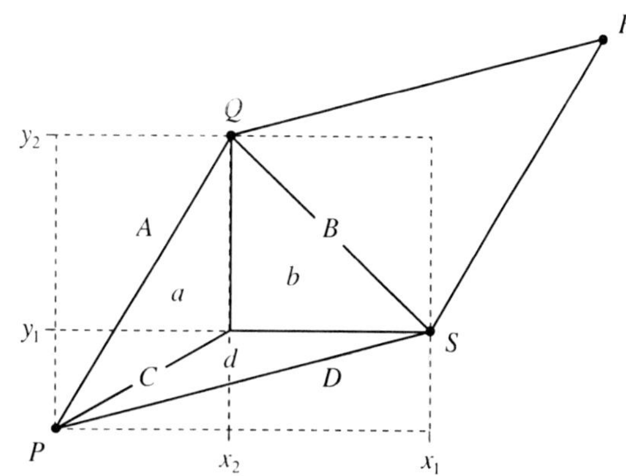

The figure shows four rectangles A, B, C, and D and three triangles a, b, and d. $x_1y_2 - x_2y_1$ is the sum of the areas of rectangles A, B, and D. Because they have the same base and height, rectangle A has twice the area of triangle a; the same is true of B and b and of D and d. Thus, quadrilateral $PQRS$ has area

$$2 \cdot [\text{area}\,(a) + \text{area}\,(b) + \text{area}\,(d)] = \text{area}\,(A) + \text{area}\,(B) + \text{area}\,(D) = x_1y_2 - x_2y_1$$

4.4.35 The n-dimensional cube has 2^n corners, $n2^{n-1}$ edges and $2n$ faces of dimension $n-1$. The cube whose edges are the rows of $2I$ has volume 2^n.

4.4.37 $J = r$. The columns are orthogonal and their lengths are 1 and r.

4.4.39 $$\begin{vmatrix} \dfrac{\partial r}{\partial x} & \dfrac{\partial r}{\partial y} \\ \dfrac{\partial \theta}{\partial x} & \dfrac{\partial \theta}{\partial y} \end{vmatrix} = \begin{vmatrix} \cos\theta & \sin\theta \\ -\dfrac{\sin\theta}{r} & \dfrac{\cos\theta}{r} \end{vmatrix} = \frac{1}{r}$$

4.4.41 $S = (2,1,-1)$ gives a parallelogram, whose area is the length of a cross product: $\|PQ \times PS\| = \|(-2,-2,-1)\| = 3$. This comes from a determinant too! The other four corners could be $(0,0,0)$, $(0,0,2)$, $(1,2,2)$, $(1,1,0)$. The volume of the tilted box is

$$\det\begin{bmatrix} PQ \\ PS \\ PO \end{bmatrix} = \begin{vmatrix} 0 & 1 & 2 \\ 1 & 2 & 2 \\ -1 & 0 & 1 \end{vmatrix} = 1.$$

4.4.43 $\det\begin{bmatrix} x & y & z \\ 3 & 2 & 1 \\ 1 & 2 & 3 \end{bmatrix} = 0 = 7x - 5y + z$; plane contains the two vectors.

4.4.45 VISA has the five reversals VI, VS, VA, IA, SA and AVIS has the two reversals VI and VS. Since $5-2$ is odd, VISA and AVIS have opposite parity.

Chapter 5 Eigenvalues and Eigenvectors

Problem Set 5.1, page 240

5.1.1 $\lambda = 2$ and $\lambda = 3$, trace $(A) = 5$, det $(A) = 6$.

5.1.3 $\lambda = -5$ and $\lambda = -4$; both λ's are reduced by 7, with unchanged eigenvectors.

5.1.5 For A: $\lambda = 3$, $\lambda = 1$, $\lambda = 0$ with eigenvectors $(1, 0, 0)$, $(2, -1, 0)$, $(3, -2, 1)$; trace $(A) = 4$, det $(A) = 0$.
For B: $\lambda = 2$, $\lambda = 2$, $\lambda = -2$ with eigenvectors $(1, 1, 1)$, $(0, 1, 0)$, $(1, 0, -1)$; trace $(B) = 2$, det $(B) = -8$.

5.1.7 $Ax = \lambda x$ gives $(A - 7I)\,x = (\lambda - 7)\,x$; $Ax = \lambda x$ gives $x = \lambda A^{-1} x$ so $A^{-1}x = (1/\lambda)\,x$.

5.1.9 The coefficient of $(-\lambda)^{n-1}$ in $(\lambda_1 - \lambda) \cdots (\lambda_n - \lambda)$ is $\lambda_1 + \cdots + \lambda_n$. In det $(A - \lambda I)$, a term which includes an off-diagonal a_{ij} excludes both $a_{ii} - \lambda$ and $a_{jj} - \lambda$. Such a term doesn't involve $(-\lambda)^{n-1}$. So the coefficient of $(-\lambda)^{n-1}$ in det $(A - \lambda I)$ must come from the product down the main diagonal. That coefficient is $a_{11} + \cdots + a_{nn} = \lambda_1 + \cdots + \lambda_n$.

5.1.11 Transpose $A - \lambda I$: $\det(A - \lambda I) = \det(A - \lambda I)^{\mathrm{T}} = \det\left(A^{\mathrm{T}} - \lambda I\right)$.

5.1.13 The eigenvalues of A are 1, 2, 3, 7, 8, 9.

5.1.15 rank $(A) = 1$, $\lambda = 0, \ldots, 0, n$ (trace n); rank $(C) = 2$, $\lambda = 0, \ldots, n/2, -n/2$ (trace 0).

5.1.17 The third row contains 6, 5, 4.

5.1.19 A and A^2 and A^∞ all have the same eigenvectors. The eigenvalues are 1 and 0.5 for A, 1 and 0.25 for A^2, 1 and 0 for A^∞. Therefore A^2 is halfway between A and A^∞.

5.1.21 $\lambda_1 = 4$ and $\lambda_2 = -1$ (check trace and determinant) with $x_1 = (1, 2)$ and $x_2 = (2, -1)$. A^{-1} has the same eigenvectors as A, with eigenvalues $1/\lambda_1 = 1/4$ and $1/\lambda_2 = -1$.

5.1.23 (a) Multiply Ax to see λx which reveals λ. (b) Solve $(A - \lambda I)\,x = 0$ to find x.

5.1.25 (a) $Pu = \left(uu^{\mathrm{T}}\right)u = u\left(u^{\mathrm{T}}u\right) = u$ so $\lambda = 1$.
(b) $Pv = \left(uu^{\mathrm{T}}\right)v = u\left(u^{\mathrm{T}}v\right) = 0$ so $\lambda = 0$.
(c) $x_1 = (-1, 1, 0, 0)$, $x_2 = (-3, 0, 1, 0)$, $x_3 = (-5, 0, 0, 1)$ are orthogonal to u, so they are eigenvectors of P with $\lambda = 0$.

5.1.27 $\lambda^3 - 1 = 0$ gives $\lambda = 1$ and $\lambda = \frac{1}{2}\left(-1 \pm i\sqrt{3}\right)$; the three eigenvalues are $1, 1, -1$.

5.1.29 (a) rank $(B) = 2$ (b) $\det\left(B^{\mathrm{T}}B\right) = 0$ (c) Not enough information (d) $(B + I)^{-1}$ has $(\lambda + 1)^{-1} = 1, \frac{1}{2}, \frac{1}{3}$.

5.1.31 $a = 0$, $b = 9$, $c = 0$ multiply $1, \lambda, \lambda^2$ in $\det(A - \lambda I) = 9\lambda - \lambda^3$: A is the *companion matrix*.

5.1.33 $\begin{bmatrix} 0 & 0 \\ 1 & 0 \end{bmatrix}, \begin{bmatrix} 0 & 1 \\ 0 & 0 \end{bmatrix}, \begin{bmatrix} -1 & 1 \\ -1 & 1 \end{bmatrix}$. Always $A^2 = 0$ if $\lambda = 0, 0$. See Problem 5.2.40 for a discussion of the Cayley-Hamilton Theorem.

5.1.35 $Ax = c_1\lambda_1 x_1 + \cdots + c_n\lambda_n x_n$ equals $Bx = c_1\lambda_1 x_1 + \cdots + c_n\lambda_n x_n$ for all x. So $A = B$.

5.1.37 $\begin{bmatrix} a & b \\ c & d \end{bmatrix}\begin{bmatrix} 1 \\ 1 \end{bmatrix} = \begin{bmatrix} a+b \\ c+d \end{bmatrix} = (a+b)\begin{bmatrix} 1 \\ 1 \end{bmatrix}$; $\lambda_2 = d - b$ in order that trace $(A) = a + d$.

5.1.39 We need $\lambda^3 = 1$ but not $\lambda = 1$ (to avoid I). With $\lambda_1 = e^{2\pi i/3}$ and $\lambda_2 = e^{-2\pi i/3}$, the determinant is $\lambda_1\lambda_2 = 1$ and the trace is $\lambda_1 + \lambda_2 = \cos\frac{2\pi}{3} + i\sin\frac{2\pi}{3} + \cos\frac{2\pi}{3} - i\sin\frac{2\pi}{3} = -1$. One matrix with trace -1 and determinant 1 is $A = \begin{bmatrix} -1 & 1 \\ -1 & 0 \end{bmatrix}$.

Problem Set 5.2, page 250

5.2.1 $\begin{bmatrix} 1 & 1 \\ 1 & 1 \end{bmatrix} = \begin{bmatrix} 1 & 1 \\ 1 & -1 \end{bmatrix}\begin{bmatrix} 2 & 0 \\ 0 & 0 \end{bmatrix}\begin{bmatrix} 1 & 1 \\ 1 & -1 \end{bmatrix}^{-1}$; $\begin{bmatrix} 2 & 1 \\ 0 & 0 \end{bmatrix} = \begin{bmatrix} 1 & 1 \\ 0 & -2 \end{bmatrix}\begin{bmatrix} 2 & 0 \\ 0 & 0 \end{bmatrix}\begin{bmatrix} 1 & 1 \\ 0 & -2 \end{bmatrix}^{-1}$.

5.2.3 $\lambda = 0, 0, 3$; the third column of S is a multiple of $\begin{bmatrix} 1 \\ 1 \\ 1 \end{bmatrix}$ and the other columns are on the plane orthogonal to it.

5.2.5 A_1 and A_3 cannot be diagonalized. They have only one line of eigenvectors.

5.2.7 $A = \begin{bmatrix} 3 & 1 \\ 1 & -1 \end{bmatrix}\begin{bmatrix} 5 & 0 \\ 0 & 1 \end{bmatrix}\begin{bmatrix} 3 & 1 \\ 1 & -1 \end{bmatrix}^{-1}$ gives $A^{100} = \begin{bmatrix} 3 & 1 \\ 1 & -1 \end{bmatrix}\begin{bmatrix} 5^{100} & 0 \\ 0 & 1 \end{bmatrix}\begin{bmatrix} 3 & 1 \\ 1 & -1 \end{bmatrix}^{-1} =$
$\frac{1}{4}\begin{bmatrix} 3\cdot 5^{100}+1 & 3\cdot 5^{100}-3 \\ 5^{100}-1 & 5^{100}+3 \end{bmatrix}$.

5.2.9 trace (AB) = trace (BA) $= aq + bs + cr + dt$. Then trace $(AB - BA) = 0$ always. So $AB - BA = I$ is impossible for matrices, since I does not have trace zero.

5.2.11 (a) True; $\det(A) = 2 \neq 0$. (b) False; $\begin{bmatrix} 1 & 1 & 1 \\ 0 & 1 & 1 \\ 0 & 0 & 2 \end{bmatrix}$ (c) False; $\begin{bmatrix} 1 & 0 & 0 \\ 0 & 1 & 0 \\ 0 & 0 & 2 \end{bmatrix}$ is diagonal!

5.2.13 $A = \begin{bmatrix} 1 & 1 \\ 1 & -1 \end{bmatrix}\begin{bmatrix} 9 & 0 \\ 0 & 1 \end{bmatrix}\begin{bmatrix} 1 & 1 \\ 1 & -1 \end{bmatrix}^{-1}$; $\pm\begin{bmatrix} 2 & 1 \\ 1 & 2 \end{bmatrix}$ or $\pm\begin{bmatrix} 1 & 2 \\ 2 & 1 \end{bmatrix}$; four square roots.

5.2.15 $\begin{bmatrix} 1 & 2 \\ 0 & 3 \end{bmatrix} = \begin{bmatrix} 1 & 1 \\ 0 & 1 \end{bmatrix}\begin{bmatrix} 1 & 0 \\ 0 & 3 \end{bmatrix}\begin{bmatrix} 1 & -1 \\ 0 & 1 \end{bmatrix}$; $\begin{bmatrix} 1 & 1 \\ 2 & 2 \end{bmatrix} = \begin{bmatrix} 1 & 1 \\ -1 & 2 \end{bmatrix}\begin{bmatrix} 0 & 0 \\ 0 & 3 \end{bmatrix}\begin{bmatrix} \frac{2}{3} & -\frac{1}{3} \\ \frac{1}{3} & \frac{1}{3} \end{bmatrix}$.

5.2.17 $A = \begin{bmatrix} 1 & 1 \\ 0 & 1 \end{bmatrix}\begin{bmatrix} 2 & 0 \\ 0 & 5 \end{bmatrix}\begin{bmatrix} 1 & -1 \\ 0 & 1 \end{bmatrix} = \begin{bmatrix} 2 & 3 \\ 0 & 5 \end{bmatrix}$.

5.2.19 (a) False: don't know λ's (b) True (c) True (d) False: need eigenvectors of S!

5.2.21 The columns of S are multiples of $(2, 1)$ and $(0, 1)$ in either order. Same for A^{-1}.

5.2.23 A and B have $\lambda_1 = 1$ and $\lambda_2 = 1$. $A + B$ has $\lambda_1 = 1, \lambda_2 = 3$. Eigenvalues of $A + B$ *are not equal* to eigenvalues of A plus eigenvalues of B.

5.2.25 (a) True (b) False (c) False (A might have 2 or 3 independent eigenvectors).

5.2.27 $A = \begin{bmatrix} 8 & 3 \\ -3 & 2 \end{bmatrix}$ (or other), $A = \begin{bmatrix} 9 & 4 \\ -4 & 1 \end{bmatrix}$, $A = \begin{bmatrix} 10 & 5 \\ -5 & 0 \end{bmatrix}$; only eigenvectors are $(c, -c)$.

5.2.29 $S\lambda^k S^{-1}$ approaches zero if and only if every $|\lambda| < 1$; $B^k \to 0$ from $\lambda = .9$ and $\lambda = .3$.

5.2.31 $\Lambda = \begin{bmatrix} 0.9 & 0 \\ 0 & 0.3 \end{bmatrix}$, $S = \begin{bmatrix} 3 & -3 \\ 1 & 1 \end{bmatrix}$; $B^{10}\begin{bmatrix} 3 \\ 1 \end{bmatrix} = (0.9)^{10}\begin{bmatrix} 3 \\ 1 \end{bmatrix}$, $B^{10}\begin{bmatrix} 3 \\ -1 \end{bmatrix} = (0.3)^{10}\begin{bmatrix} 3 \\ -1 \end{bmatrix}$,
$B^{10}\begin{bmatrix} 6 \\ 0 \end{bmatrix} = B^{10}\begin{bmatrix} 3 \\ 1 \end{bmatrix} + B^{10}\begin{bmatrix} 3 \\ -1 \end{bmatrix}$.

5.2.33 $B^k = \begin{bmatrix} 1 & 1 \\ 0 & -1 \end{bmatrix}\begin{bmatrix} 3 & 0 \\ 0 & 2 \end{bmatrix}^k\begin{bmatrix} 1 & 1 \\ 0 & -1 \end{bmatrix} = \begin{bmatrix} 3^k & 3^k - 2^k \\ 0 & 2^k \end{bmatrix}$.

5.2.35 trace $(AB) = (aq + bs) + (cr + dt) = (qa + rc) + (sb + td) =$ trace (BA). Proof for diagonalizable case: the trace of $S\Lambda S^{-1}$ is the trace of $(\Lambda S^{-1})\, S = \Lambda$ which is *the sum of the* λ's.

5.2.37 The A's form a subspace since cA and $A_1 + A_2$ have the same S. When $S = I$ the A's give the subspace of diagonal matrices. Its dimension is 4.

5.2.39 Two problems: The nullspace and column space can overlap, so x could be in both. There may not be r independent eigenvectors in the column space.

5.2.41 $A = \begin{bmatrix} 1 & 1 \\ 1 & 0 \end{bmatrix}$ has $A^2 = \begin{bmatrix} 2 & 1 \\ 1 & 1 \end{bmatrix}$ and $A^2 - A - I = 0$ confirms Cayley-Hamilton.

5.2.43 By Theorem 5F, B has the same eigenvectors $(1, 0)$ and $(0, 1)$ as A, so B is also diagonal. The equations $AB - BA = \begin{bmatrix} a & b \\ 2c & 2d \end{bmatrix} - \begin{bmatrix} a & 2b \\ c & 2d \end{bmatrix} = \begin{bmatrix} 0 & 0 \\ 0 & 0 \end{bmatrix}$ are $-b = 0$ and $c = 0$: rank 2.

5.2.45 A has $\lambda_1 = 1$ and $\lambda_2 = .4$ with $x_1 = (1, 2)$ and $x_2 = (1, -1)$. A^∞ has $\lambda_1 = 1$ and $\lambda_2 = 0$ (same eigenvectors). A^{100} has $\lambda_1 = 1$ and $\lambda_2 = (0.4)^{100}$ which is near zero. So A^{100} is very near A^∞.

Problem Set 5.3, page 262

5.3.1 The Fibonacci numbers start even, odd, odd. Then *odd* + *odd* = *even*. The next two are odd (from *odd* + *even* and *even* + *odd*). Then repeat *odd* + *odd* = *even*.

5.3.3 $A^2 = \begin{bmatrix} 2 & 1 \\ 1 & 1 \end{bmatrix}$, $A^3 = \begin{bmatrix} 3 & 2 \\ 2 & 1 \end{bmatrix}$, $A^4 = \begin{bmatrix} 5 & 3 \\ 3 & 2 \end{bmatrix}$; $F_{20} = 6765$.

5.3.5 $A = S\Lambda S^{-1} = \begin{bmatrix} 1 & 1 \\ 1 & 0 \end{bmatrix} = \dfrac{1}{\lambda_1 - \lambda_2}\begin{bmatrix} \lambda_1 & \lambda_2 \\ 1 & 1 \end{bmatrix}\begin{bmatrix} \lambda_1 & 0 \\ 0 & \lambda_2 \end{bmatrix}\begin{bmatrix} 1 & -\lambda_2 \\ -1 & \lambda_1 \end{bmatrix}$ (notice S^{-1}).

$$S\Lambda^k S^{-1} = \frac{1}{\lambda_1 - \lambda_2}\begin{bmatrix} \lambda_1 & \lambda_2 \\ 1 & 1 \end{bmatrix}\begin{bmatrix} \lambda_1^k & 0 \\ 0 & \lambda_2^k \end{bmatrix}\begin{bmatrix} 1 & -\lambda_2 \\ -1 & \lambda_1 \end{bmatrix}\begin{bmatrix} 1 \\ 0 \end{bmatrix} = \begin{bmatrix} \cdots \\ \dfrac{\lambda_1^k - \lambda_2^k}{\lambda_1 - \lambda_2} \end{bmatrix}.$$

5.3.7 Direct addition $L_k + L_{k+1}$ gives $L_0, \ldots, L_{10}$ as $2, 1, 3, 4, 7, 11, 18, 29, 47, 76, 123$.
A calculator gives $\lambda_1^{10} = (1.618\ldots)^{10} = 122.991\ldots$ which rounds off to $L_{10} = 123$.

5.3.9 The Markov transition matrix is $\begin{bmatrix} \frac{7}{12} & \frac{1}{6} & 0 \\ \frac{1}{6} & \frac{1}{2} & 0 \\ \frac{1}{4} & \frac{1}{3} & 1 \end{bmatrix}$. Fractions $\frac{7}{12}, \frac{1}{2}, 1$ don't move.

5.3.11 (a) $\lambda = 0, (1, 1, -2)$ (b) $\lambda = 1$ and -0.2 (c) The limit $(3, 4, 4)$ is the eigenvector corresponding to $\lambda = 1$.

5.3.13 (a) $0 \le a \le 1, 0 \le b \le 1$

(b) $u_k = \begin{bmatrix} \frac{b}{1-a} & 1 \\ 1 & -1 \end{bmatrix} \begin{bmatrix} 1^k & 0 \\ 0 & (a-b)^k \end{bmatrix} \begin{bmatrix} \frac{b}{1-a} & 1 \\ 1 & -1 \end{bmatrix}^{-1} \begin{bmatrix} 1 \\ 1 \end{bmatrix}$

$= \begin{bmatrix} \frac{2b}{b-a+1} - \frac{1-a-b}{b-a+1}(a-b)^k \\ \frac{2(1-a)}{b-a+1} - \frac{1-a-b}{b-a+1}(a-b)^k \end{bmatrix}$

(c) $u_k \to \begin{bmatrix} \frac{2b}{b-a+1} \\ \frac{2(1-a)}{b-a+1} \end{bmatrix}$ if $|a-b| < 1$. If $a = \frac{1}{3}$ and $b = -\frac{1}{3}$ then the matrix is not Markov.

5.3.15 The components of Ax add to $x_1 + x_2 + x_3$ (each column adds to 1 and nobody is lost). The components of λx add to $\lambda(x_1 + x_2 + x_3)$. If $\lambda \neq 1$, $x_1 + x_2 + x_3$ must be zero.

5.3.17 $\begin{bmatrix} \alpha & \alpha \\ \alpha & \alpha \end{bmatrix}$ is unstable for $|\alpha| > \frac{1}{2}$, and stable for $|\alpha| < \frac{1}{2}$. It is neutral for $\alpha = \pm\frac{1}{2}$.

5.3.19 $A^2 = \begin{bmatrix} 0 & 0 & 2 \\ 0 & 0 & 0 \\ 0 & 0 & 0 \end{bmatrix}$ and $A^3 = 0$. So $(I-A)^{-1} = I + A + A^2 = \begin{bmatrix} 1 & 1 & 2 \\ 0 & 1 & 1 \\ 0 & 0 & 1 \end{bmatrix}$.

5.3.21 If A is increased, then more goods are consumed in production and the expansion must be slower. Mathematically, $Ax \geq tx$ remains true if A is increased; $t_{\max}$ goes up.

5.3.23 $\begin{bmatrix} 3 & 2 \\ 2 & 3 \end{bmatrix} = \frac{1}{2}\begin{bmatrix} 1 & -1 \\ 1 & 1 \end{bmatrix}\begin{bmatrix} 5 & 0 \\ 0 & 1 \end{bmatrix}\begin{bmatrix} 1 & 1 \\ -1 & 1 \end{bmatrix}$ and $A^k = \frac{1}{2}\begin{bmatrix} 1 & -1 \\ 1 & 1 \end{bmatrix}\begin{bmatrix} 5^k & 0 \\ 0 & 1 \end{bmatrix}\begin{bmatrix} 1 & 1 \\ -1 & 1 \end{bmatrix}$.

5.3.25 $R = S\sqrt{\Lambda}S^{-1} = \begin{bmatrix} 2 & 1 \\ 1 & 2 \end{bmatrix}$ has $R^2 = A$. $\sqrt{B}$ would have $\lambda = \sqrt{9}$ and $\lambda = \sqrt{-1}$ so its trace is not real. Note $\begin{bmatrix} -1 & 0 \\ 0 & -1 \end{bmatrix}$ can have $\sqrt{-1} = i$ and $-i$, and real square root $\begin{bmatrix} 0 & 1 \\ -1 & 0 \end{bmatrix}$.

5.3.27 $A = S\lambda_1 S^{-1}$ and $B = S\lambda_2 S^{-1}$. Diagonal matrices always give $\lambda_1\lambda_2 = \lambda_2\lambda_1$. Then $AB = BA$ from $S\lambda_1 S^{-1} S\lambda_2 S^{-1} = S\lambda_1\lambda_2 S^{-1} = S\lambda_2\lambda_1 S^{-1} = S\lambda_2 S^{-1} S\lambda_1 S^{-1} = BA$.

5.3.29 B has $\lambda = i$ and $-i$, so B^4 has $\lambda^4 = 1$ and 1; C has $\lambda = \frac{1}{2}(1 \pm \sqrt{3}i) = e^{\pm\pi i/3}$ so $\lambda^3 = -1$ and -1. Then $C^3 = -I$ and $C^{1024} = -C$.

Problem Set 5.4, page 275

5.4.1 $\lambda_1 = -2$ and $\lambda_2 = 0$; $x_1 = (1,-1)$ and $x_2 = (1,1)$; $e^{At} = \frac{1}{2}\begin{bmatrix} e^{-2t}+1 & -e^{-2t}+1 \\ -e^{-2t}+1 & e^{-2t}+1 \end{bmatrix}$.

5.4.3 $u(t) = \begin{bmatrix} e^{2t}+2 \\ -e^{2t}+2 \end{bmatrix}$; as $t \to \infty$, $e^{2t} \to \infty$.

5.4.5 (a) $e^{A(t+T)} = Se^{\lambda(t+T)}S^{-1} = Se^{\lambda t}e^{\lambda T}S^{-1} = Se^{\lambda t}S^{-1}Se^{\lambda T}S^{-1} = e^{At}e^{AT}$

(b) $e^A = I + A = \begin{bmatrix} 1 & 0 \\ 1 & 1 \end{bmatrix}$, $e^B = I + B = \begin{bmatrix} 1 & -1 \\ 0 & 1 \end{bmatrix}$, $A + B = \begin{bmatrix} 0 & -1 \\ 1 & 0 \end{bmatrix}$ gives $e^{A+B} = \begin{bmatrix} \cos 1 & -\sin 1 \\ \sin 1 & \cos 1 \end{bmatrix}$ from Example 1 in the text, at $t = 1$. This matrix is different from $e^A e^B$.

5.4.7 $e^{At} = I + At = \begin{bmatrix} 1 & t \\ 0 & 1 \end{bmatrix}$; $e^{\lambda t} u(0) = \begin{bmatrix} 4t + 3 \\ 4 \end{bmatrix}$.

5.4.9 (a) $\lambda_1 = \frac{7+\sqrt{57}}{2}$, $\lambda_2 = \frac{7-\sqrt{57}}{2}$, $\operatorname{Re} \lambda_1 > 0$, unstable.

(b) $\lambda_1 = \sqrt{7}$, $\lambda_2 = -\sqrt{7}$, $\operatorname{Re} \lambda_1 > 0$, unstable.

(c) $\lambda_1 = \frac{-1+\sqrt{13}}{2}$, $\lambda_2 = \frac{-1-\sqrt{13}}{2}$, $\operatorname{Re} \lambda_1 > 0$, unstable.

(d) $\lambda_1 = 0$, $\lambda_2 = -2$, neutrally stable.

5.4.11 A_1 is unstable for $t < 1$, neutrally stable for $t \geq 1$. A_2 is unstable for $t < 4$, neutrally stable at $t = 4$, stable with real λ for $4 < t \leq 5$, and stable with complex λ for $t > 5$. A_3 is unstable for all $t > 0$ because the trace is $2t$.

5.4.13 (a) $u_1' = cu_2 - bu_3$, $u_2' = -cu_1 + au_3$, $u_3' = bu_1 - au_2$ gives $u_1' u_1 + u_2' u_2 + u_3' u_3 = 0$.

(b) Because e^{At} is an orthogonal matrix, $\|u(t)\|^2 = \left\|e^{At} u(0)\right\|^2 = \|u(0)\|^2$ is constant.

(c) $\lambda = 0$ and $\pm \left(\sqrt{a^2 + b^2 + c^2}\right) i$. Skew-symmetric matrices have pure imaginary λ's.

5.4.15 $u(t) = \frac{1}{2} \cos 2t \begin{bmatrix} 1 \\ -1 \end{bmatrix} + \frac{1}{2} \cos \sqrt{6} t \begin{bmatrix} 1 \\ 1 \end{bmatrix}$.

5.4.17 $Ax = \lambda F x + \lambda^2 x$ or $\left(A - \lambda F - \lambda^2 I\right) x = 0$.

5.4.19 Eigenvalues are real when $(\text{trace } A)^2 - 4 \det A \geq 0 \Rightarrow -4\left(-a^2 - b^2 + c^2\right) \geq 0 \Rightarrow a^2 + b^2 \geq c^2$.

5.4.21 $u_1 = e^{4t} \begin{bmatrix} 1 \\ 0 \end{bmatrix}$, $u_2 = e^t \begin{bmatrix} 1 \\ -1 \end{bmatrix}$. If $u(0) = (5, -2)$, then $u(t) = 3e^{4t} \begin{bmatrix} 1 \\ 0 \end{bmatrix} + 2e^t \begin{bmatrix} 1 \\ -1 \end{bmatrix}$.

5.4.23 $\begin{bmatrix} y' \\ y'' \end{bmatrix} = \begin{bmatrix} 0 & 1 \\ 4 & 5 \end{bmatrix} \begin{bmatrix} y \\ y' \end{bmatrix}$. Then $\lambda = \frac{1}{2}\left(5 \pm \sqrt{41}\right)$.

5.4.25 $\lambda_1 = 0$ and $\lambda_2 = 2$. Now $v(t) = 20 + 10e^{2t} \to \infty$ as $t \to \infty$.

5.4.27 $A = \begin{bmatrix} 0 & 1 \\ -9 & 6 \end{bmatrix}$ has trace 6, determinant 9, $\lambda = 3$ and 3 with only one independent eigenvector $(1, 3)$. That gives $y = ce^{3t}$, $y' = 3e^{3t}$. Also te^{3t} solves $y'' = 6y' - 9y$.

5.4.29 (a) $y(t) = \cos t$ starts at $y(0) = 1$ and $y'(0) = 0$. $y(t) = \sin t$ also solves $d^2 y / dt^2 = -y$.

(b) The vector equation has $u = (y, y') = (\cos t, -\sin t)$.

5.4.31 Substituting $u = e^{ct} v$ gives $ce^{ct} v = Ae^{ct} v - e^{ct} b$ or $(A - cI) v = b$ or $v = (A - cI)^{-1} b$, the particular solution. If c is an eigenvalue then $A - cI$ is not invertible: this v fails.

5.4.33 $de^{At}/dt = A + A^2 t + \frac{1}{2} A^3 t^2 + \frac{1}{6} A^4 t^3 + \cdots = A\left(I + At + \frac{1}{2} A^2 t^2 + \frac{1}{6} A^3 t^3 + \cdots\right) = Ae^{At}$.

5.4.35 The solution at time $t + T$ is also $e^{A(t+T)} u(0)$. Thus $e^{At} \cdot e^{AT} = e^{A(t+T)}$.

5.4.37 If $A^2 = A$ then $e^{At} = I + At + \frac{1}{2} At^2 + \frac{1}{6} At^3 + \cdots = I + \left(e^t - 1\right) A$ (as in Problem 4). Continuing, this is equal to $\begin{bmatrix} 1 & 0 \\ 0 & 1 \end{bmatrix} + \begin{bmatrix} e^t - 1 & e^t - 1 \\ 0 & 0 \end{bmatrix} = \begin{bmatrix} e^t & e^t - 1 \\ 0 & 1 \end{bmatrix}$.

5.4.39 $A = \begin{bmatrix} 1 & 1 \\ 0 & 3 \end{bmatrix} = \begin{bmatrix} 1 & 1 \\ 2 & 0 \end{bmatrix} \begin{bmatrix} 3 & 0 \\ 0 & 1 \end{bmatrix} \begin{bmatrix} 0 & \frac{1}{2} \\ 1 & -\frac{1}{2} \end{bmatrix}$. Then $e^{At} = \begin{bmatrix} e^t & \frac{1}{2}(e^{3t} - e^t) \\ 0 & e^{3t} \end{bmatrix} = I$ at $t = 0$.

5.4.41 (a) The inverse of e^{At} is e^{-At}. (b) If $Ax = \lambda x$, then $e^{At}x = e^{\lambda t}x$ and $e^{\lambda t} \neq 0$.

5.4.43 $\lambda = 2$ and 5 with eigenvectors $\begin{bmatrix} 2 \\ 1 \end{bmatrix}$ and $\begin{bmatrix} 1 \\ 1 \end{bmatrix}$. Then $A = S\lambda S^{-1} = \begin{bmatrix} -1 & 6 \\ -3 & 8 \end{bmatrix}$.

Problem Set 5.5, page 288

5.5.1 (a)

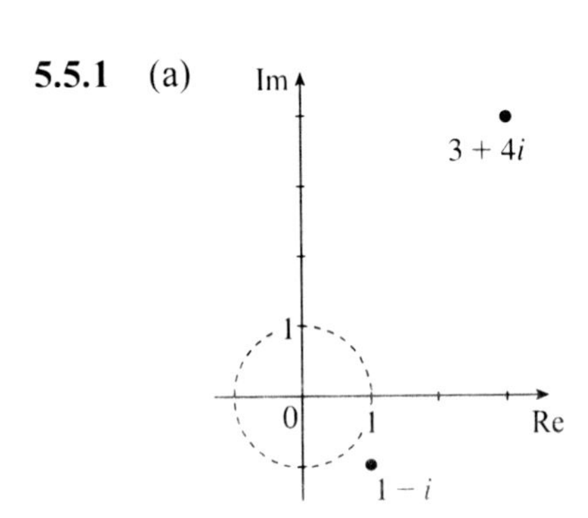

(b) sum $4 + 3i$; product $7 + i$

(c) $\overline{3 + 4i} = 3 - 4i$; $\overline{1 - i} = 1 + i$; $|3 + 4i| = 5$; $|1 - i| = \sqrt{2}$. Both numbers lie *outside* the unit circle.

5.5.3 $\overline{x} = 2 - i$, $x\,\|x\| = 5$, $xy = -1 + 7i$, $1/x = \frac{2}{5} - \frac{1}{5}i$, $x/y = \frac{1}{2} - \frac{1}{2}i$; check that $|xy| = \sqrt{50} = |x|\,|y|$ and $|1/x| = \frac{1}{\sqrt{5}} = 1/\,|x|$.

5.5.5 (a) $x^2 = r^2 e^{i2\theta}$, $x^{-1} = (1/r)\,e^{-i\theta}$, $\overline{x} = re^{-i\theta}$; $x^{-1} = \overline{x}$ gives $|x|^2 = 1$: on the unit circle.

(b)

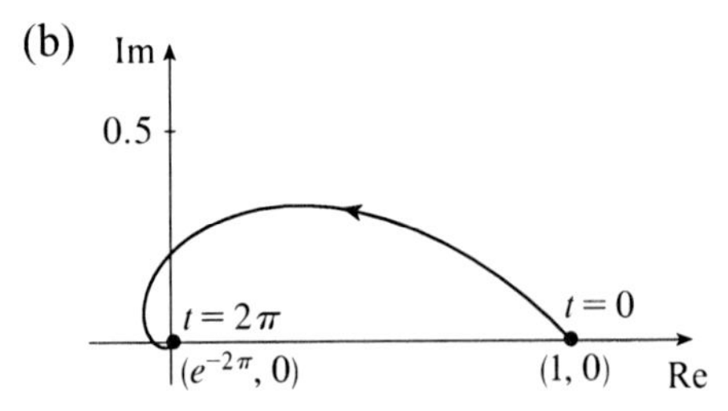

5.5.7 $C = \begin{bmatrix} 1 & -i \\ -i & 0 \\ 0 & 1 \end{bmatrix} \begin{bmatrix} 1 & i & 0 \\ i & 0 & 1 \end{bmatrix} = \begin{bmatrix} 2 & i & -i \\ -i & 1 & 0 \\ i & 0 & 1 \end{bmatrix}$, $C^{\mathrm{H}} = C$ because $\left(A^{\mathrm{H}}A\right)^{\mathrm{H}} = A^{\mathrm{H}}A$.

5.5.9 (a) $\det A^{\mathrm{T}} = \det A$ but $\det A^{\mathrm{H}} = \overline{\det A}$. (b) $A^{\mathrm{H}} = A$ gives $\overline{\det A} = \det A$, which is therefore real.

5.5.11 P: $\lambda_1 = 0$, $\lambda_2 = 1$, $x_1 = \begin{bmatrix} \frac{1}{\sqrt{2}} \\ -\frac{1}{\sqrt{2}} \end{bmatrix}$, $x_2 = \begin{bmatrix} \frac{1}{\sqrt{2}} \\ \frac{1}{\sqrt{2}} \end{bmatrix}$; Q: $\lambda_1 = 1$, $\lambda_2 = -1$, $x_1 = \begin{bmatrix} \frac{1}{\sqrt{2}} \\ \frac{1}{\sqrt{2}} \end{bmatrix}$, $x_2 = \begin{bmatrix} \frac{1}{\sqrt{2}} \\ -\frac{1}{\sqrt{2}} \end{bmatrix}$; R: $\lambda_1 = 5$, $\lambda_2 = -5$, $x_1 = \begin{bmatrix} \frac{2}{\sqrt{5}} \\ \frac{1}{\sqrt{5}} \end{bmatrix}$, $x_2 = \begin{bmatrix} \frac{1}{\sqrt{5}} \\ -\frac{2}{\sqrt{5}} \end{bmatrix}$.

5.5.13 (a) u, v, w are mutually orthogonal.

(b) The nullspace is spanned by u; the left nullspace is the same as the nullspace; the row space is spanned by v and w; the column space is the same as the row space.

(c) $x = v + \frac{1}{2}w$; not unique, we can add any multiple of u to x.

(d) Need $b^{\mathrm{T}}u = 0$.

(e) $S^{-1} = S^{\mathrm{T}}$; $S^{-1}AS = \mathrm{diag}\,(0, 1, 2)$.

5.5.15 The dimension of S is $\dfrac{n(n+1)}{2}$, not n. Every symmetric matrix A is a combination of n projections, but the projections change as A changes. There is no basis of n fixed projection matrices, in the space $\mathbf{S}$ of symmetric matrices.

5.5.17 $(UV)^{\mathrm{H}}(UV) = V^{\mathrm{H}}U^{\mathrm{H}}UV = V^{\mathrm{H}}IV = I$. So UV is unitary.

5.5.19 The third column of U can be $\frac{1}{\sqrt{6}}(1, -2, i)$, multiplied by any number $e^{i\theta}$.

5.5.21 A has ± 1 in each diagonal entry; eight possibilities.

5.5.23 Columns of Fourier matrix U are eigenvectors of P because $PU = \operatorname{diag}\left(1, w, w^2, w^3\right)U$ (and $w = i$).

5.5.25 n^2 steps for direct C times x; only $n \log n$ steps for F and F^{-1} by FFT (and n for λ).

5.5.27 $A^{\mathrm{H}}A = \begin{bmatrix} 2 & 0 & 1+i \\ 0 & 2 & 1+i \\ 1-i & 1-i & 2 \end{bmatrix}$ and $AA^{\mathrm{H}} = \begin{bmatrix} 3 & 1 \\ 1 & 3 \end{bmatrix}$ are *Hermitian* matrices.

$\left(A^{\mathrm{H}}A\right)^{\mathrm{H}} = A^{\mathrm{H}}A^{\mathrm{HH}} = A^{\mathrm{H}}A$ again.

5.5.29 cA is still Hermitian for real c; $(iA)^{\mathrm{H}} = -iA^{\mathrm{H}} = -iA$ is skew-Hermitian.

5.5.31 $P^2 = \begin{bmatrix} 0 & 0 & 1 \\ 1 & 0 & 0 \\ 0 & 1 & 0 \end{bmatrix}$, $P^3 = I$, $P^{100} = P^{99}P = P$; the λ are the cube roots of 1: 1, $e^{2\pi i/3}$, $e^{4\pi i/3}$.

5.5.33 $C = \begin{bmatrix} 2 & 5 & 4 \\ 4 & 2 & 5 \\ 5 & 4 & 2 \end{bmatrix} = 2 + 5P + 4P^2$ has $\lambda(C) = \left\{\begin{matrix} 2+5+4 \\ 2+5e^{2\pi i/3}+4e^{4\pi i/3} \\ 2+5e^{4\pi i/3}+4e^{8\pi i/3} \end{matrix}\right\}$.

5.5.35 $A = \dfrac{1}{\sqrt{3}}\begin{bmatrix} 1 & -1+i \\ 1+i & 1 \end{bmatrix}\begin{bmatrix} 2 & 0 \\ 0 & -1 \end{bmatrix}\dfrac{1}{\sqrt{3}}\begin{bmatrix} 1 & 1-i \\ -1-i & 1 \end{bmatrix}$. $K = \left(iA^{\mathrm{T}}\right) =$ $\dfrac{1}{\sqrt{3}}\begin{bmatrix} 1 & -1-i \\ 1-i & 1 \end{bmatrix}\begin{bmatrix} 2i & 0 \\ 0 & -i \end{bmatrix}\dfrac{1}{\sqrt{3}}\begin{bmatrix} 1 & 1+i \\ -1+i & 1 \end{bmatrix}$.

5.5.37 $V = \dfrac{1}{L}\begin{bmatrix} 1+\sqrt{3} & -1+i \\ 1+i & 1+\sqrt{3} \end{bmatrix}\begin{bmatrix} 1 & 0 \\ 0 & -1 \end{bmatrix}\dfrac{1}{L}\begin{bmatrix} 1+\sqrt{3} & 1-i \\ -1-i & 1+\sqrt{3} \end{bmatrix}$ with $L^2 = 6 + 2\sqrt{3}$. $V = V^{\mathrm{H}}$ gives real λ, unitary gives $|\lambda| = 1$, then trace zero gives $\lambda = 1, -1$.

5.5.39 Don't multiply e^{-ix} times e^{ix}; conjugate the first, then $\int_0^{2\pi} e^{2ix}\,dx = \left[e^{2ix}/2i\right]_0^{2\pi} = 0$.

5.5.41 $R + iS = (R + iS)^{\mathrm{H}} = R^{\mathrm{T}} - iS^{\mathrm{T}}$; R is symmetric but S is skew-symmetric.

5.5.43 $\begin{bmatrix} 1 \end{bmatrix}$ and $\begin{bmatrix} -1 \end{bmatrix}$; $\begin{bmatrix} a & b+ic \\ b-ic & -a \end{bmatrix}$ with $a^2 + b^2 + c^2 = 1$.

5.5.45 $\left(I - 2uu^{\mathrm{H}}\right)^{\mathrm{H}} = I - 2uu^{\mathrm{H}}$; $\left(I - 2uu^{\mathrm{H}}\right)^2 = I - 4uu^{\mathrm{H}} + 4u\left(u^{\mathrm{H}}u\right)u^{\mathrm{H}} = I$; the matrix uu^{H} projects onto the line through u.

5.5.47 We are given $A + iB = (A + iB)^{\mathrm{H}} = A^{\mathrm{T}} - iB^{\mathrm{T}}$. Then $A = A^{\mathrm{T}}$ and $B = -B^{\mathrm{T}}$.

5.5.49 $A = \begin{bmatrix} 1-i & 1-i \\ -1 & 2 \end{bmatrix}\begin{bmatrix} 1 & 0 \\ 0 & 4 \end{bmatrix}\dfrac{1}{6}\begin{bmatrix} 2+2i & -2 \\ 1+i & 2 \end{bmatrix} = S\Lambda S^{-1}$. A has real eigenvalues 1 and 4.

Problem Set 5.6, page 302

5.6.1 $C = N^{-1}BN = N^{-1}M^{-1}AMN = (MN)^{-1}A(MN)$; only $M^{-1}IM = I$ is similar to I.

5.6.3 If $\lambda_1, \ldots, \lambda_n$ are eigenvalues of A, then $\lambda_1 + 1, \ldots, \lambda_n + 1$ are eigenvalues of $A + I$. So A and $A + I$ never have the same eigenvalues, and can't be similar.

5.6.5 If B is invertible then $BA = B(AB)B^{-1}$ is similar to AB.

5.6.7 The $(3,1)$ entry of $M^{-1}AM$ is $g\cos\theta + h\sin\theta$, which is zero if $\tan\theta = -g/h$.

5.6.9 The coefficients are $c_1 = 1$, $c_2 = 2$, $d_1 = 1$, $d_2 = 1$; check $Mc = d$.

5.6.11 The reflection matrix with basis v_1 and v_2 is $A = \begin{bmatrix} 0 & 1 \\ 1 & 0 \end{bmatrix}$. The basis V_1 and V_2 (same reflection!) gives $B = \begin{bmatrix} 1 & 0 \\ 0 & -1 \end{bmatrix}$. If $M = \begin{bmatrix} 1 & 1 \\ 1 & -1 \end{bmatrix}$ then $A = MBM^{-1}$.

5.6.13 (a) $D = \begin{bmatrix} 0 & 1 & 0 \\ 0 & 0 & 2 \\ 0 & 0 & 0 \end{bmatrix}$

(b) $D^3 = \begin{bmatrix} 0 & 0 & 0 \\ 0 & 0 & 0 \\ 0 & 0 & 0 \end{bmatrix}$, the third derivative matrix. The third derivatives of 1, x and x^2 are zero, so $D^3 = 0$.

(c) $\lambda = 0$ (triple); only one independent eigenvector $(1, 0, 0)$.

5.6.15 The eigenvalues are $1, 1, 1, -1$. Eigenmatrices $\begin{bmatrix} 1 & 0 \\ 0 & 0 \end{bmatrix}, \begin{bmatrix} 0 & 1 \\ 1 & 0 \end{bmatrix}, \begin{bmatrix} 0 & 0 \\ 0 & 1 \end{bmatrix}, \begin{bmatrix} 0 & 1 \\ -1 & 0 \end{bmatrix}$.

5.6.17 (i) $TT^{\mathrm{H}} = U^{-1}AUU^{\mathrm{H}}A^{\mathrm{H}}(U^{-1})^{\mathrm{H}} = I$.

(ii) If T is triangular and unitary, then its diagonal entries (the eigenvalues) must have absolute value 1. Then all off-diagonal entries are 0 because the columns are to be unit vectors.

5.6.19 The $(1,1)$ entries of $T^{\mathrm{H}}T = TT^{\mathrm{H}}$ give $|t_{11}|^2 = |t_{11}|^2 + |t_{12}|^2 + |t_{13}|^2$ so $t_{12} = t_{13} = 0$. Comparing the $(2,2)$ entries of $T^{\mathrm{H}}T = TT^{\mathrm{H}}$ gives $t_{23} = 0$. So T must be diagonal.

5.6.21 If $N = U\lambda U^{-1}$ then $NN^{\mathrm{H}} = U\lambda U^{-1}(U^{-1})^{\mathrm{H}}\lambda^{\mathrm{H}}U^{\mathrm{H}}$ is equal to $U\lambda\lambda^{\mathrm{H}}U^{\mathrm{H}}$. This is the same as $U\lambda^{\mathrm{H}}\lambda U^{\mathrm{H}} = (U\lambda U^{-1})^{\mathrm{H}}(U\lambda U^{-1}) = N^{\mathrm{H}}N$. So N is normal.

5.6.23 The eigenvalues of $A(A - I)(A - 2I)$ are $0, 0, 0$.

5.6.25 Always $\begin{bmatrix} a^2 + bc & ab + bd \\ ac + cd & bc + d^2 \end{bmatrix} - (a + d)\begin{bmatrix} a & b \\ c & d \end{bmatrix} + (ad - bc)\begin{bmatrix} 1 & 0 \\ 0 & 1 \end{bmatrix} = \begin{bmatrix} 0 & 0 \\ 0 & 0 \end{bmatrix}$!

5.6.27 $M^{-1}J_3M = 0$, so the last two inequalities are easy. Trying for $MJ_1 = J_2M$ forces the first column of M to be zero, so M cannot be invertible. Cannot have $J_1 = M^{-1}J_2M$.

5.6.29 $A^{10} = 2^{10}\begin{bmatrix} 61 & 45 \\ -80 & -59 \end{bmatrix}$; $e^A = e^2\begin{bmatrix} 13 & 9 \\ -16 & -11 \end{bmatrix}$.

5.6.31 $\begin{bmatrix} 1 & 1 \\ 0 & 0 \end{bmatrix}, \begin{bmatrix} 0 & 0 \\ 1 & 1 \end{bmatrix}, \begin{bmatrix} 1 & 0 \\ 1 & 0 \end{bmatrix}, \begin{bmatrix} 0 & 1 \\ 0 & 1 \end{bmatrix}$ are similar with eigenvalues 0, 1; $\begin{bmatrix} 1 & 0 \\ 0 & 1 \end{bmatrix}$ has eigenvalues 1, 1; $\begin{bmatrix} 0 & 1 \\ 1 & 0 \end{bmatrix}$ has eigenvalues $1, -1$.

5.6.33 (a) $(M^{-1}AM)(M^{-1}x) = M^{-1}(Ax) = M^{-1}0 = 0$

(b) The nullspaces of A and of $M^{-1}AM$ have the same *dimension*. Different vectors, different bases.

5.6.35 $J^2 = \begin{bmatrix} c^2 & 2c \\ 0 & c^2 \end{bmatrix}$, $J^3 = \begin{bmatrix} c^3 & 3c^2 \\ 0 & c^3 \end{bmatrix}$, $J^k = \begin{bmatrix} c^k & kc^{k-1} \\ 0 & c^k \end{bmatrix}$; $J^0 = I$, $J^{-1} = \begin{bmatrix} c^{-1} & -c^{-2} \\ 0 & c^{-1} \end{bmatrix}$.

5.6.37 $w(t) = \left(w(0) + tx(0) + \frac{1}{2}t^2 y(0) + \frac{1}{6}t^3 z(0)\right) e^{5t}$.

5.6.39 (a) Choose M_i to be the reverse diagonal matrix to get $M_i^{-1} J_i M_i = M_i^{\mathrm{T}}$ in each block.

(b) M_0 has those blocks M_i on its diagonal, so $M_0^{-1} J M_0 = J^{\mathrm{T}}$.

(c) $A^{\mathrm{T}} = (M^{-1})^{\mathrm{T}} J^{\mathrm{T}} M^{\mathrm{T}}$ is $(M^{-1})^{\mathrm{T}} M_0^{-1} J M_0 M^{\mathrm{T}} = (M M_0 M^{\mathrm{T}})^{-1} A (M M_0 M^{\mathrm{T}})$, and A^{T} is similar to A.

5.6.41 (a) True: One has $\lambda = 0$, the other doesn't.

(b) False. Diagonalize a nonsymmetric matrix and λ is symmetric.

(c) False: $\begin{bmatrix} 0 & 1 \\ -1 & 0 \end{bmatrix}$ and $\begin{bmatrix} 0 & -1 \\ 1 & 0 \end{bmatrix}$ are similar.

(d) True: All eigenvalues of $A + I$ are increased by 1, and thus differ from the eigenvalues of A.

5.6.43 (a) The diagonal blocks are 6 by 6 and 4 by 4 (b) AB has all the same eigenvalues as BA plus $6 - 4 = 2$ zeros.

Chapter 6 Positive Definite Matrices

Problem Set 6.1, page 316

6.1.1 $ac - b^2 = 2 - 4 = -2 < 0$; $x^2 + 4xy + 2y^2 = (x + 2y)^2 - 2y^2$ (difference of squares).

6.1.3 $\det(A - \lambda I) = \lambda^2 - (a + c)\lambda + ac - b^2 = 0$ gives $\lambda_1 = \frac{1}{2}\left((a + c) + \sqrt{(a - c)^2 + b^2}\right)$ and $\lambda_2 = \frac{1}{2}\left((a + c) - \sqrt{(a - c)^2 + 4b^2}\right)$; $\lambda_1 > 0$ is a sum of positive numbers; $\lambda_2 > 0$ because $(a + c)^2 > (a - c)^2 + 4b^2$ reduces to $ac > b^2$. Better way: product $\lambda_1\lambda_2 = ac - b^2$.

6.1.5 (a) Positive definite when $-3 < b < 3$.

(b) $\begin{bmatrix} 1 & b \\ b & 9 \end{bmatrix} = \begin{bmatrix} 1 & 0 \\ b & 1 \end{bmatrix}\begin{bmatrix} 1 & 0 \\ 0 & 9 - b^2 \end{bmatrix}\begin{bmatrix} 1 & b \\ 0 & 1 \end{bmatrix}$

(c) The minimum is $-\dfrac{1}{2(9 - b^2)}$ when $\begin{bmatrix} 1 & b \\ b & 9 \end{bmatrix}\begin{bmatrix} x \\ y \end{bmatrix} = \begin{bmatrix} 0 \\ 1 \end{bmatrix}$ which is $\begin{bmatrix} x \\ y \end{bmatrix} = \dfrac{1}{9 - b^2}\begin{bmatrix} -b \\ 1 \end{bmatrix}$.

(d) There is no minimum. If $y \to \infty$ along the line $x = -3y$, then $x - y$ approaches $-\infty$.

6.1.7 (a) $A_1 = \begin{bmatrix} 1 & -1 & -1 \\ -1 & 1 & 1 \\ -1 & 1 & 1 \end{bmatrix}$ and $A_2 = \begin{bmatrix} 1 & -1 & -1 \\ -1 & 2 & -2 \\ -1 & -2 & 11 \end{bmatrix}$.

(b) $f_1 = (x_1 - x_2 - x_3)^2 = 0$ when $x_1 - x_2 - x_3 = 0$.

(c) $f_2 = (x_1 - x_2 - x_3)^2 + (x_2 - 3x_3)^2 + x_3^2$; $L = \begin{bmatrix} 1 & 0 & 0 \\ -1 & 1 & 0 \\ -1 & -3 & 1 \end{bmatrix}$.

6.1.9 $A = \begin{bmatrix} 3 & 6 \\ 6 & 16 \end{bmatrix} = \begin{bmatrix} 1 & 0 \\ 2 & 1 \end{bmatrix}\begin{bmatrix} 3 & 0 \\ 0 & 4 \end{bmatrix}\begin{bmatrix} 1 & 2 \\ 0 & 1 \end{bmatrix}$; the coefficients of the squares are the pivots in D, while the coefficients inside the squares are columns of L.

6.1.11 (a) Pivots are a and $c - |b|^2/a$ and $\det A = ac - |b|^2$.

(c) Multiply $|x_2|^2$ by $\left(c - |b|^2/a\right)$.

(c) Now $x^{\mathrm{H}}Ax$ is a sum of squares.

(d) $\det\begin{bmatrix} 1 & 1 + i \\ 1 - i & 2 \end{bmatrix} = 0$ (indefinite) and $\det\begin{bmatrix} 3 & 4 + i \\ 4 - i & 6 \end{bmatrix} = 1$ (positive definite).

6.1.13 $a > 1$ and $(a - 1)(c - 1) > b^2$. This means that $A - I$ is positive definite.

6.1.15 $f(x, y) = x^2 + 4xy + 9y^2 = (x + 2y)^2 + 5y^2$; $f(x, y) = x^2 + 6xy + 9y^2 = (x + 3y)^2$.

6.1.17 $x^{\mathrm{T}}A^{\mathrm{T}}Ax = (Ax)^{\mathrm{T}}(Ax) =$ length squared, which is 0 only if $Ax = 0$. Since A has independent columns this happens only when $x = 0$.

6.1.19 $A = \begin{bmatrix} 4 & -4 & 8 \\ -4 & 4 & -8 \\ 8 & -8 & 16 \end{bmatrix}$ has only one pivot $= 4$, rank $A = 1$, eigenvalues 24, 0, 0, and $\det A = 0$.

6.1.21 $ax^2 + 2bxy + cy^2$ has a saddle point at $(0, 0)$ if $ac < b^2$. The matrix is *indefinite* ($\lambda < 0$ and $\lambda > 0$).

Problem Set 6.2, page 326

6.2.1 A is positive definite for $a > 2$. B is never positive definite: notice $\begin{bmatrix} 1 & 4 \\ 4 & 7 \end{bmatrix}$.

6.2.3 $\det A = -2b^3 - 3b^2 + 1$ is negative at (and near) $b = \frac{2}{3}$.

6.2.5 If $x^{\mathrm{T}}Ax > 0$ and $x^{\mathrm{T}}Bx > 0$ for any $x \neq 0$, then $x^{\mathrm{T}}(A+B)x > 0$; condition I.

6.2.7 Positive λ's because R is symmetric and $\sqrt{\lambda} > 0$. $R = \begin{bmatrix} 3 & 1 \\ 1 & 3 \end{bmatrix}$; $R = \begin{bmatrix} 3 & -1 \\ -1 & 3 \end{bmatrix}$.

6.2.9 $|x^{\mathrm{T}}Ay|^2 = |x^{\mathrm{T}}R^{\mathrm{T}}Ry|^2 = \left|(Rx)^{\mathrm{T}}Ry\right|^2 \leq \|Rx\|^2\|Ry\|^2$ by the ordinary Schwarz inequality. Then $\|Rx\|^2\|Ry\|^2 = (x^{\mathrm{T}}R^{\mathrm{T}}Rx)(y^{\mathrm{T}}R^{\mathrm{T}}Ry) = (x^{\mathrm{T}}Ax)(y^{\mathrm{T}}Ay)$.

6.2.11 $A = \begin{bmatrix} 3 & -\sqrt{2} \\ -\sqrt{2} & 2 \end{bmatrix}$ has $\lambda = 1$ and 4, axes $1\begin{bmatrix} 1 \\ \sqrt{2} \end{bmatrix}$ and $\frac{1}{2}\begin{bmatrix} \sqrt{2} \\ -1 \end{bmatrix}$ along eigenvectors.

6.2.13 *Negative definite matrices*:

(I) $x^{\mathrm{T}}Ax < 0$ for all nonzero vectors x.

(II) All the eigenvalues of A satisfy $\lambda_i < 0$.

(III) $\det A_1 < 0, \det A_2 > 0, \det A_3 < 0$.

(IV) All the pivots (without row exchanges) satisfy $d_i < 0$.

(V) There is a matrix R with independent columns such that $A = -R^{\mathrm{T}}R$.

6.2.15 (a) False (Q must contain eigenvectors of A)

(b) True (same eigenvalues as A)

(c) True ($Q^{\mathrm{T}}AQ = Q^{-1}AQ$ is similar to A)

(d) True (eigenvalues of e^{-A} are $e^{-\lambda} > 0$)

6.2.17 Start from $a_{jj} = (\text{row } j \text{ of } R^{\mathrm{T}})(\text{column } j \text{ of } R) =$ length squared of column j of R. Then

$$\det A = (\det R)^2 = (\text{volume of the } R \text{ parallelepiped})^2 \leq \begin{array}{c}\text{product of the lengths squared} \\ \text{of all the columns of } R\end{array} = a_{11}a_{22}\cdots a_{nn}$$

6.2.19 $A = \begin{bmatrix} 2 & -1 & 0 \\ -1 & 2 & -1 \\ 0 & -1 & 2 \end{bmatrix}$ has pivots $2, \frac{3}{2}, \frac{4}{3}$; $A = \begin{bmatrix} 2 & -1 & -1 \\ -1 & 2 & -1 \\ -1 & -1 & 2 \end{bmatrix}$ is singular; $A\begin{bmatrix} 1 \\ 1 \\ 1 \end{bmatrix} = \begin{bmatrix} 0 \\ 0 \\ 0 \end{bmatrix}$.

6.2.21 $x^{\mathrm{T}}Ax$ is not positive when $(x_1, x_2, x_3) = (0, 1, 0)$ because of the zero on the diagonal.

6.2.23 (a) Positive definite requires positive determinant (also: all $\lambda > 0$).

(b) All projection matrices except I are singular.

(c) The diagonal entries of D are its eigenvalues.

(d) The negative definite matrix $-I$ has determinant 1 when n is even.

6.2.25 $\lambda_1 = 1/a^2$ and $\lambda_2 = 1/b^2$ so $a = 1/\sqrt{\lambda_1}$ and $b = 1/\sqrt{\lambda_2}$. The ellipse $9x^2 + 16y^2 = 1$ has axes with half-lengths $a = \frac{1}{3}$ and $b = \frac{1}{4}$.

6.2.27 $A = \begin{bmatrix} 9 & 3 \\ 3 & 5 \end{bmatrix} = \begin{bmatrix} 3 & 0 \\ 1 & 2 \end{bmatrix}\begin{bmatrix} 3 & 1 \\ 0 & 2 \end{bmatrix}$; $C = \begin{bmatrix} 2 & 0 \\ 4 & 3 \end{bmatrix}$ has $CC^{\mathrm{T}} = \begin{bmatrix} 4 & 8 \\ 8 & 25 \end{bmatrix}$.

6.2.29 $ax^2 + 2bxy + cy^2 = a\left(x + \frac{b}{a}y\right)^2 + \frac{ac-b^2}{a}y^2$; $2x^2 + 8xy + 10y^2 = 2(x+2y)^2 + 2y^2$.

6.2.31 $x^{\mathrm{T}}Ax = 2\left(x_1 - \frac{1}{2}x_2 - \frac{1}{2}x_3\right)^2 + \frac{3}{2}(x_2 - x_3)^2$; $x^{\mathrm{T}}Bx = (x_1 + x_2 + x_3)^2$. B has one pivot.

6.2.33 A and $C^{\mathrm{T}}AC$ have $\lambda_1 > 0$, $\lambda_2 = 0$. $C(t) = tQ + (1-t)QR$, $Q = \begin{bmatrix} 1 & 0 \\ 0 & -1 \end{bmatrix}$, $R = \begin{bmatrix} 2 & 0 \\ 0 & 1 \end{bmatrix}$; C has one positive and one negative eigenvalue, but I has two positive eigenvalues.

6.2.35 The pivots of $A - \frac{1}{2}I$ are 2.5, 5.9, -0.81, so one eigenvalue of $A - \frac{1}{2}I$ is negative. Then A has an eigenvalue smaller than $\frac{1}{2}$.

6.2.37 $\operatorname{rank}(C^{\mathrm{T}}AC) \le \operatorname{rank} A$, but also $\operatorname{rank}(C^{\mathrm{T}}AC) \ge \operatorname{rank}\left((C^{\mathrm{T}})^{-1}C^{\mathrm{T}}ACC^{-1}\right) = \operatorname{rank} A$.

6.2.39 No. If C is not square, $C^{\mathrm{T}}AC$ is not the same size as A.

6.2.41 $\det\begin{bmatrix} 6 - \frac{4}{18}\lambda & -3 - \frac{1}{18}\lambda \\ -3 - \frac{1}{18}\lambda & 6 - \frac{4}{18}\lambda \end{bmatrix} = 0$ gives $\lambda_1 = 54$, $\lambda_2 = \frac{54}{5}$. Eigenvectors $\begin{bmatrix} 1 \\ -1 \end{bmatrix}, \begin{bmatrix} 1 \\ 1 \end{bmatrix}$.

6.2.43 *Groups*: orthogonal matrices Q; all exponentials e^{tA} of a fixed matrix A; matrices D with determinant 1. *Not Groups:* Positive definite symmetric matrices A (not closed under multiplication); matrices P with positive eigenvalues (not closed under multiplication). If A is positive definite, the group of all powers A^k contains only positive definite matrices.

Problem Set 6.3, page 337

6.3.1 $A^{\mathrm{T}}A = \begin{bmatrix} 5 & 20 \\ 20 & 80 \end{bmatrix}$ has only $\sigma_1^2 = 85$ with $v_1 = \begin{bmatrix} \frac{1}{\sqrt{17}} \\ \frac{4}{\sqrt{17}} \end{bmatrix}$ so $v_2 = \begin{bmatrix} \frac{4}{\sqrt{17}} \\ -\frac{1}{\sqrt{17}} \end{bmatrix}$.

6.3.3 $A^{\mathrm{T}}A = \begin{bmatrix} 2 & 1 \\ 1 & 1 \end{bmatrix}$ has eigenvalues $\sigma_1^2 = \dfrac{3+\sqrt{5}}{2}$ and $\sigma_2^2 = \dfrac{3-\sqrt{5}}{2}$.

Since $A = A^{\mathrm{T}}$ the eigenvectors of $A^{\mathrm{T}}A$ are the same as for A. Since $\lambda_2 = \frac{1}{2}(1-\sqrt{5})$ is *negative*, $\sigma_1 = \lambda_1$ but $\sigma_2 = -\lambda_2$. The unit eigenvectors are the same as in Section 6.2 for A, except for the effect of this minus sign (because we need $Av_2 = \sigma_2 u_2$):

$$u_1 = v_1 = \begin{bmatrix} \dfrac{\lambda_1}{\sqrt{1+\lambda_1^2}} \\ \dfrac{1}{\sqrt{1+\lambda_1^2}} \end{bmatrix} \text{ and } u_2 = -v_2 = \begin{bmatrix} \dfrac{\lambda_2}{\sqrt{1+\lambda_2^2}} \\ \dfrac{1}{\sqrt{1+\lambda_2^2}} \end{bmatrix}.$$

6.3.5 $AA^{\mathrm{T}} = \begin{bmatrix} 2 & 1 \\ 1 & 2 \end{bmatrix}$ has $\sigma_1^2 = 3$ with $u_1 = \begin{bmatrix} \frac{1}{\sqrt{2}} \\ \frac{1}{\sqrt{2}} \end{bmatrix}$ and $\sigma_2^2 = 1$ with $u_2 = \begin{bmatrix} \frac{1}{\sqrt{2}} \\ -\frac{1}{\sqrt{2}} \end{bmatrix}$.

$A^{\mathrm{T}}A = \begin{bmatrix} 1 & 1 & 0 \\ 1 & 2 & 1 \\ 0 & 1 & 1 \end{bmatrix}$ has $\sigma_1^2 = 3$ with $v_1 = \begin{bmatrix} \frac{1}{\sqrt{6}} \\ \frac{2}{\sqrt{6}} \\ \frac{1}{\sqrt{6}} \end{bmatrix}$, $\sigma_2^2 = 1$ with $v_2 = \begin{bmatrix} \frac{1}{\sqrt{2}} \\ 0 \\ -\frac{1}{\sqrt{2}} \end{bmatrix}$; and nullvector

$v_3 = \begin{bmatrix} \frac{1}{\sqrt{3}} \\ -\frac{1}{\sqrt{3}} \\ \frac{1}{\sqrt{3}} \end{bmatrix}$. Then $\begin{bmatrix} 1 & 1 & 0 \\ 0 & 1 & 1 \end{bmatrix} = \begin{bmatrix} u_1 & u_2 \end{bmatrix} \begin{bmatrix} \sqrt{3} & 0 & 0 \\ 0 & 1 & 0 \end{bmatrix} \begin{bmatrix} v_1 & v_2 & v_3 \end{bmatrix}^{\mathrm{T}}$.

6.3.7 $A = 12\,uv^{\mathrm{T}}$ has one singular value $\sigma_1 = 12$.

6.3.9 Multiply $U\Sigma V^{\mathrm{T}}$ using columns (of U) times rows (of ΣV^{T}).

6.3.11 To make A singular, the smallest change sets its smallest singular value σ_2 to zero.

6.3.13 The singular values of $A+I$ are not $\sigma_j + 1$. They come from eigenvalues of $(A+I)^{\mathrm{T}}(A+I)$.

6.3.15 $A^+ = \begin{bmatrix} \frac{1}{4} \\ \frac{1}{4} \\ \frac{1}{4} \\ \frac{1}{4} \end{bmatrix}$, $B = \begin{bmatrix} 0 & 1 \\ 1 & 0 \end{bmatrix}\begin{bmatrix} 1 & 0 & 0 \\ 0 & 1 & 0 \end{bmatrix}\begin{bmatrix} 1 & 0 & 0 \\ 0 & 1 & 0 \\ 0 & 0 & 1 \end{bmatrix}$, $B^+ = \begin{bmatrix} 0 & 1 \\ 1 & 0 \\ 0 & 0 \end{bmatrix}$, $C^+ = \begin{bmatrix} \frac{1}{2} & 0 \\ \frac{1}{2} & 0 \end{bmatrix}$.

A^+ is the right-inverse of A; B^+ is the left-inverse of B.

6.3.17 $A^{\mathrm{T}}A = \begin{bmatrix} 10 & 6 \\ 6 & 10 \end{bmatrix} = \frac{1}{2}\begin{bmatrix} 1 & -1 \\ 1 & 1 \end{bmatrix}\begin{bmatrix} 4 & 0 \\ 0 & 16 \end{bmatrix}\begin{bmatrix} 1 & 1 \\ -1 & 1 \end{bmatrix}$, take square roots of 4 and 16 to obtain

$$S = \frac{1}{2}\begin{bmatrix} 1 & -1 \\ 1 & 1 \end{bmatrix}\begin{bmatrix} 2 & 0 \\ 0 & 4 \end{bmatrix}\begin{bmatrix} 1 & 1 \\ -1 & 1 \end{bmatrix} = \begin{bmatrix} 3 & 1 \\ 1 & 3 \end{bmatrix} \text{ and } Q = AS^{-1} = \frac{1}{\sqrt{10}}\begin{bmatrix} 3 & 1 \\ -1 & 3 \end{bmatrix}.$$

6.3.19 (a) With independent columns, the row space is all of $\mathbf{R}^n$; check $\left(A^{\mathrm{T}}A\right)A^+b = A^{\mathrm{T}}b$.

(b) $A^{\mathrm{T}}\left(AA^{\mathrm{T}}\right)^{-1}b$ is in the row space because A^{T} times any vector is in that space; now $\left(A^{\mathrm{T}}A\right)A^+b = A^{\mathrm{T}}AA^{\mathrm{T}}\left(AA^{\mathrm{T}}\right)^{-1}b = A^{\mathrm{T}}b$. Both cases give $A^{\mathrm{T}}Ax^+ = A^{\mathrm{T}}b$.

6.3.21 Take $A = \begin{bmatrix} 1 & 1 \\ 0 & 0 \end{bmatrix}$ and $B = \begin{bmatrix} 0 & 0 \\ 1 & 1 \end{bmatrix}$. Then $AB = \begin{bmatrix} 1 & 1 \\ 0 & 0 \end{bmatrix}$. From C^+ in Problem 15 we have

$$A^+ = \begin{bmatrix} \frac{1}{2} & 0 \\ \frac{1}{2} & 0 \end{bmatrix} \text{ and } B^+ = \begin{bmatrix} 0 & \frac{1}{2} \\ 0 & \frac{1}{2} \end{bmatrix} = (AB)^+ \text{ and } (AB)^+ \neq B^+A^+.$$

6.3.23 $A = Q_1\Sigma Q_2^{\mathrm{T}} \Rightarrow A^+ = Q_2\Sigma^+Q_1^{\mathrm{T}} \Rightarrow AA^+ = Q_1\Sigma\Sigma^+Q_1^{\mathrm{T}}$. Squaring gives $(AA^+)^2 = Q_1\Sigma\Sigma^+\Sigma\Sigma^+Q_1^{\mathrm{T}} = Q_1\Sigma\Sigma^+Q_1^{\mathrm{T}}$. So we have projections: $(AA^+)^2 = AA^+ = (AA^+)^{\mathrm{T}}$ and similarly for A^+A. AA^+ and A^+A project onto the column space and row space of A.

Problem Set 6.4, page 344

6.4.1 $P(x) = x_1^2 - x_1x_2 + x_2^2 - x_2x_3 + x_3^2 - 4x_1 - 4x_3$ has $\partial P/\partial x_1 = 2x_1 - x_2 - 4$ and $\partial P/\partial x_2 = -x_1 + 2x_2 - x_3$ and $\partial P/\partial x_3 = -x_2 + 2x_3 - 4$.

6.4.3 $\partial P_1/\partial x = x + y = 0$ and $\partial P_1/\partial y = x + 2y - 3 = 0$ give $x = -3$ and $y = 3$. P_2 has no minimum (let $y \to \infty$). It is associated with the semidefinite matrix $\begin{bmatrix} 1 & 0 \\ 0 & 0 \end{bmatrix}$.

6.4.5 Put $x = (1, \ldots, 1)$ in Rayleigh's quotient (the denominator becomes n). Since $R(x)$ is always between λ_1 and λ_n we get $n\lambda_1 \le x^{\mathrm{T}}Ax = \sum a_{ij} \le n\lambda_n$.

6.4.7 Since $x^{\mathrm{T}}Bx > 0$ for all nonzero vectors x, $x^{\mathrm{T}}(A+B)x$ will be larger than $x^{\mathrm{T}}Ax$. So the Rayleigh quotient is larger for $A+B$ (in fact all n eigenvalues are increased).

6.4.9 Since $x^{\mathrm{T}}Bx > 0$, the Rayleigh quotient for $A+B$ is larger than the quotient for A.

6.4.11 The smallest eigenvalues in $Ax = \lambda x$ and $Ax = \lambda Mx$ are $\frac{1}{2}$ and $\frac{1}{4}\left(3 - \sqrt{3}\right)$.

6.4.13 (a) $\lambda_j = \min\limits_{S_j}\left[\max\limits_{x \text{ in } S_j} R(x)\right] > 0$ means that every S_j contains a vector x with $R(x) > 0$.

(b) $y = C^{-1}x$ gives quotient $\overline{R}(y) = \dfrac{y^{\mathrm{T}}C^{\mathrm{T}}ACy}{y^{\mathrm{T}}y} = \dfrac{x^{\mathrm{T}}Ax}{x^{\mathrm{T}}x} = R(x) > 0.$

6.4.15 The extreme subspace S_2 is spanned by the eigenvectors x_1 and x_2.

6.4.17 If $Cx = C\left(A^{-1}b\right)$ equals d then $CA^{-1}b - d = 0$ in the correction term in (5).

Problem Set 6.5, page 350

6.5.1 $Ay = b$ is $4\begin{bmatrix} 2 & -1 & 0 \\ -1 & 2 & -1 \\ 0 & -1 & 2 \end{bmatrix}\begin{bmatrix} \frac{3}{16} \\ \frac{4}{16} \\ \frac{3}{16} \end{bmatrix} = b = \begin{bmatrix} \frac{1}{2} \\ \frac{1}{2} \\ \frac{1}{2} \end{bmatrix}$. The linear finite element $U = \frac{3}{16}V_1 + \frac{4}{16}V_2 + \frac{3}{16}V_3$ equals the exact $u = \frac{3}{16}, \frac{4}{16}, \frac{3}{16}$ at the nodes $x = \frac{1}{4}, \frac{1}{2}, \frac{3}{4}$.

6.5.3 $A_{33} = 3$, $b_3 = \frac{1}{3}$. Then $A = 3\begin{bmatrix} 2 & -1 & 0 \\ -1 & 2 & -1 \\ 0 & -1 & 1 \end{bmatrix}$, $b = \dfrac{1}{3}\begin{bmatrix} 2 \\ 2 \\ 1 \end{bmatrix}$, $Ay = b$ gives $y = \dfrac{1}{9}\begin{bmatrix} 5 \\ 8 \\ 9 \end{bmatrix}$.

6.5.5 Integrate by parts: $\int_0^1 (-V_i''V_j)\,dx = \int_0^1 V_i'V_j'\,dx - [V_i'V_j]_{x=0}^{x=1} = \int_0^1 V_i'V_j'\,dx =$ same A_{ij}.

6.5.7 $A = 4$, $M = \frac{1}{3}$. Their ratio 12 (Rayleigh quotient on the subspace of multiples of $V(x)$) is *larger* than the true eigenvalue $\lambda = \pi^2$.

6.5.9 The mass matrix M is $h/6$ times the tridiagonal 1, 4, 1 matrix.

Chapter 7 Computations with Matrices

Problem Set 7.2, page 357

7.2.1 If Q is orthogonal, its norm is $\|Q\| = \max \|Qx\| / \|x\| = 1$ because Q preserves length: $\|Qx\| = \|x\|$ for every x. Also Q^{-1} is orthogonal and has norm 1, so $c(Q) = 1$.

7.2.3 $\|ABx\| \le \|A\| \|Bx\|$, by the definition of the norm of A, and then $\|Bx\| \le \|B\| \|x\|$. Dividing by $\|x\|$ and maximizing, $\|AB\| \le \|A\| \|B\|$. The same is true for the inverse, $\|B^{-1}A^{-1}\| \le \|B^{-1}\| \|A^{-1}\|$; $c(AB) \le c(A)\, c(B)$ by multiplying these inequalities.

7.2.5 In the definition $\|A\| = \max \|Ax\|/\|x\|$, choose x to be the particular eigenvector in question; $\|Ax\| = |\lambda| \|x\|$, so the ratio is $|\lambda|$ and *maximum* ratio is *at least* $|\lambda|$.

7.2.7 $A^{\mathrm{T}}A$ and AA^{T} have the same eigenvalues since $A^{\mathrm{T}}Ax = \lambda x$ gives $AA^{\mathrm{T}}(Ax) = A\left(A^{\mathrm{T}}Ax\right) = \lambda(Ax)$. Equality of the largest eigenvalues means $\|A\| = \|A^{\mathrm{T}}\|$.

7.2.9 $A = \begin{bmatrix} 0 & 1 \\ 0 & 0 \end{bmatrix}$, $B = \begin{bmatrix} 0 & 0 \\ 1 & 0 \end{bmatrix}$, $\lambda_{\max}(A+B) > \lambda_{\max}(A) + \lambda_{\max}(B)$, (since $1 > 0 + 0$) and also $\lambda_{\max}(AB) > \lambda_{\max}(A)\,\lambda_{\max}(B)$. So $\lambda_{\max}(A)$ is not a norm.

7.2.11 (a) Yes, $c(A) = \|A\| \|A^{-1}\| = c\left(A^{-1}\right)$ since $\left(A^{-1}\right)^{-1}$ is A again.

(b) $A^{-1}b = x$ leads to $\dfrac{\|\delta b\|}{\|b\|} \le \|A\| \|A^{-1}\| \dfrac{\|\delta x\|}{\|x\|}$. This is $\dfrac{\|\delta x\|}{\|x\|} \ge \dfrac{1}{c}\dfrac{\|\delta b\|}{\|b\|}$.

7.2.13 $\|A\| = 2$ and $c = 1$; $\|A\| = \sqrt{2}$ and c is infinite (singular!); $\|A\| = \sqrt{2}$ and $c = 1$.

7.2.15 If $\lambda_{\max} = \lambda_{\min} = 1$ then all $\lambda_i = 1$ and $A = SIS^{-1} = I$. The only matrices with $\|A\| = \|A^{-1}\| = 1$ are *orthogonal matrices*, because $A^{\mathrm{T}}A$ has to be I.

7.2.17 The residual $b - Ay = \left(10^{-7}, 0\right)$ is much smaller than $b - Az = (.0013, .0016)$. But z is much closer to the solution than y.

7.2.19 $x_1^2 + \cdots + x_n^2$ is not smaller than $\max\left(x_i^2\right) = \left(\|x\|_\infty\right)^2$ and not larger than $(|x_1| + \cdots + |x_n|)^2$ which is $\left(\|x\|_1\right)^2$. Certainly $x_1^2 + \cdots + x_n^2 \le n \max\left(x_i^2\right)$ so $\|x\| \le \sqrt{n}\,\|x\|_\infty$. Choose $y = (\operatorname{sgn}(x_1), \operatorname{sgn}(x_2), \ldots, \operatorname{sgn}(x_n))$ to get $x \cdot y = \|x\|_1$. By Schwarz this is at most $\|x\| \|y\| = \sqrt{n}\,\|x\|$. Choose $x = (1, 1, \ldots, 1)$ for maximum ratios $\sqrt{n}$.

7.2.21 The exact inverse of the 3 by 3 Hilbert matrix is $A^{-1} = \begin{bmatrix} 9 & -36 & 30 \\ -36 & 192 & -180 \\ 30 & -180 & 180 \end{bmatrix}$.

7.2.23 The largest $\|x\| = \|A^{-1}b\|$ is $1/\lambda_{\min}$; the largest error is $10^{-16}/\lambda_{\min}$.

7.2.25 Exchange $\begin{bmatrix} 1 & 0 \\ 2 & 2 \end{bmatrix}$ to $\begin{bmatrix} 2 & 2 \\ 1 & 0 \end{bmatrix} \to \begin{bmatrix} 2 & 2 \\ 0 & -1 \end{bmatrix} = U$ with $P = \begin{bmatrix} 0 & 1 \\ 1 & 0 \end{bmatrix}$ and $L = \begin{bmatrix} 1 & 0 \\ .5 & 1 \end{bmatrix}$.

$$A \to \begin{bmatrix} 2 & 2 & 0 \\ 1 & 0 & 1 \\ 0 & 2 & 0 \end{bmatrix} \to \begin{bmatrix} 2 & 2 & 0 \\ 0 & -1 & 1 \\ 0 & 2 & 0 \end{bmatrix} \to \begin{bmatrix} 2 & 2 & 0 \\ 0 & 2 & 0 \\ 0 & -1 & 1 \end{bmatrix} \to \begin{bmatrix} 2 & 2 & 0 \\ 0 & 2 & 0 \\ 0 & 0 & 1 \end{bmatrix} = U.$$

Then $PA = LU$ with $P = \begin{bmatrix} 0 & 1 & 0 \\ 0 & 0 & 1 \\ 1 & 0 & 0 \end{bmatrix}$ and $L = \begin{bmatrix} 1 & 0 & 0 \\ 0 & 1 & 0 \\ .5 & -.5 & 1 \end{bmatrix}$.

Problem Set 7.3, page 365

7.3.1 $u_0 = \begin{bmatrix} 1 \\ 0 \end{bmatrix}$, $u_1 = \begin{bmatrix} 2 \\ -1 \end{bmatrix}$, $u_2 = \begin{bmatrix} 5 \\ -4 \end{bmatrix}$, $u_3 = \begin{bmatrix} 14 \\ -13 \end{bmatrix}$; $u_\infty = \frac{1}{\sqrt{2}} \begin{bmatrix} 1 \\ -1 \end{bmatrix}$ normalized to unit vector

7.3.3 $\frac{u_k}{\lambda_1^k} = c_1 x_1 + c_2 x_2 \left(\frac{\lambda_2}{\lambda_1}\right)^k + \cdots + c_n x_n \left(\frac{\lambda_n}{\lambda_1}\right)^k \to c_1 x_1$ if all ratios $\left|\frac{\lambda_i}{\lambda_1}\right| < 1$. The largest ratio controls, when k is large. $A = \begin{bmatrix} 0 & 1 \\ 1 & 0 \end{bmatrix}$ has $|\lambda_2| = |\lambda_1|$ and no convergence.

7.3.5 $Hx = x - (x-y) \frac{2(x-y)^{\mathrm{T}} x}{(x-y)^{\mathrm{T}} (x-y)} = x - (x-y) = y$. Then $H(Hx) = Hy$ is $x = Hy$.

7.3.7 $U = \begin{bmatrix} 1 & 0 \\ 0 & H \end{bmatrix} \begin{bmatrix} 1 & 0 & 0 \\ 0 & -\frac{3}{5} & -\frac{4}{5} \\ 0 & -\frac{4}{5} & \frac{3}{5} \end{bmatrix} = U^{-1}$ and then $U^{-1}AU = \begin{bmatrix} 1 & -5 & 0 \\ -5 & \frac{9}{25} & \frac{12}{25} \\ 0 & \frac{12}{25} & \frac{16}{25} \end{bmatrix}$.

7.3.9 $\begin{bmatrix} \cos\theta & \sin\theta \\ \sin\theta & 0 \end{bmatrix} = QR = \begin{bmatrix} \cos\theta & -\sin\theta \\ \sin\theta & \cos\theta \end{bmatrix} \begin{bmatrix} 1 & \cos\theta\sin\theta \\ 0 & -\sin^2\theta \end{bmatrix}$. Then $RQ = \begin{bmatrix} c(1+s^2) & -s^3 \\ -s^3 & -s^2 c \end{bmatrix}$.

7.3.11 *Assume* that $(Q_0 \cdots Q_{k-1})(R_{k-1} \cdots R_0)$ is the QR factorization of A^k (certainly true if $k = 1$). By construction $A_{k+1} = R_k Q_k$, so $R_k = A_{k+1} Q_k^{\mathrm{T}} = (Q_k^{\mathrm{T}} \cdots Q_0^{\mathrm{T}} A Q_0 \cdots Q_k) Q_k^{\mathrm{T}}$.
Postmultiplying by $R_{k-1} \cdots R_0$, the assumption gives $R_k \cdots R_0 = Q_k^{\mathrm{T}} \cdots Q_0^{\mathrm{T}} A^{k+1}$. After moving the Q's to the left side, this is the required result for A^{k+1}.

7.3.13 A has eigenvalues 4 and 2. Put one unit eigenvector in row 1 of P: it is either $\frac{1}{\sqrt{2}} \begin{bmatrix} 1 & -1 \\ 1 & 1 \end{bmatrix}$ and $PAP^{-1} = \begin{bmatrix} 2 & -4 \\ 0 & 4 \end{bmatrix}$ or $\frac{1}{\sqrt{10}} \begin{bmatrix} 1 & -3 \\ 3 & 1 \end{bmatrix}$ and $PAP^{-1} = \begin{bmatrix} 4 & -4 \\ 0 & 2 \end{bmatrix}$.

7.3.15 $P_{ij}A$ uses $4n$ multiplications (2 for each entry in rows i and j). By factoring out $\cos\theta$, the entries 1 and $\pm\tan\theta$ need only $2n$ multiplications, which leads to $\frac{2}{3}n^3$ for PR.

Problem Set 7.4, page 372

7.4.1 $D^{-1}(-L-U) = \begin{bmatrix} 0 & \frac{1}{2} & 0 \\ \frac{1}{2} & 0 & \frac{1}{2} \\ 0 & \frac{1}{2} & 0 \end{bmatrix}$, eigenvalues $\mu = 0, \pm\frac{1}{\sqrt{2}}$; $(D+L)^{-1}(-U) = \begin{bmatrix} 1 & \frac{1}{2} & 0 \\ 0 & \frac{1}{4} & \frac{1}{2} \\ 0 & \frac{1}{8} & \frac{1}{4} \end{bmatrix}$, eigenvalues $0, 0, \frac{1}{2}$; $\omega_{\text{opt}} = 4 - 2\sqrt{2}$, reducing $\lambda_{\max}$ to $3 - 2\sqrt{2} \approx 0.2$.

7.4.3 $Ax_k = (2 - 2\cos k\pi h) x_k$; $Jx_k = \frac{1}{2}(\sin 2k\pi h, \sin 3k\pi h + \sin k\pi h, \ldots) = (\cos k\pi h) x_k$. For $h = \frac{1}{4}$, A has eigenvalues $2 - 2\cos\frac{\pi}{4} = 2 - \sqrt{2}$, $2 - \cos\frac{\pi}{2} = 2$, $2 - \cos\frac{3\pi}{4} = 2 + \sqrt{2}$.

7.4.5 $J = D^{-1}(L+U) = -\begin{bmatrix} 0 & \frac{1}{3} & \frac{1}{3} \\ 0 & 0 & \frac{1}{4} \\ \frac{2}{5} & \frac{2}{5} & 0 \end{bmatrix}$; the three circles have radius $r_1 = \frac{2}{3}$, $r_2 = \frac{1}{4}$, $r_3 = \frac{4}{5}$. Their centers are at zero, so all $|\lambda_i| \le \frac{4}{5} < 1$.

7.4.7 $-D^{-1}(L+U) = \begin{bmatrix} 0 & -b/a \\ -c/d & 0 \end{bmatrix}$ has $\mu = \pm\left(\frac{bc}{ad}\right)^{1/2}$; $-(D+L)^{-1}U = \begin{bmatrix} 0 & -b/a \\ 0 & bc/(ad) \end{bmatrix}$, $\lambda = 0, \frac{bc}{ad}$; $\lambda_{\max}$ does equal $\mu^2_{\max}$.

7.4.9 If $Ax = \lambda x$ then $(I - A)x = (1 - \lambda)x$. Real eigenvalues of $B = I - A$ have $|1 - \lambda| < 1$ provided λ is between 0 and 2.

7.4.11 Always $\|AB\| \le \|A\|\,\|B\|$. Choose $A = B$ to find $\|B^2\| \le \|B\|^2$. Then choose $A = B^2$ to find $\|B^3\| \le \|B^2\|\,\|B\| \le \|B\|^3$. Continue (or use induction). Since $\|B\| \ge \max|\lambda(B)|$ it is no surprise that $\|B\| < 1$ gives convergence.

7.4.13 Jacobi has $S^{-1}T = \frac{1}{3}\begin{bmatrix} 0 & 1 \\ 1 & 0 \end{bmatrix}$ with $|\lambda|_{\max} = \frac{1}{3}$. Gauss-Seidel has $S^{-1}T = \begin{bmatrix} 0 & \frac{1}{3} \\ 0 & \frac{1}{9} \end{bmatrix}$ with $|\lambda|_{\max} = \frac{1}{9} = (|\lambda|_{\max} \text{ for Jacobi})^2$.

7.4.15 Successive overrelaxation (SOR) to be executed in MATLAB.

7.4.17 The maximum row sums give all $|\lambda| \le .9$ and $|\lambda| \le 4$. The circles around diagonal entries give tighter bounds. First A: The circle $|\lambda - .2| \le .7$ contains the other circles $|\lambda - .3| \le .5$ and $|\lambda - .1| \le .6$ and all three eigenvalues. Second A: The circle $|\lambda - 2| \le 2$ contains the circle $|\lambda - 2| \le 1$ and all three eigenvalues $2 + \sqrt{2}$, 2, and $2 - \sqrt{2}$.

7.4.19 $r_1 = b - \alpha_1 Ab = b - (b^{\mathrm{T}}b/b^{\mathrm{T}}Ab)\,Ab$ is orthogonal to $r_0 = b$: *the residuals* $r = b - Ax$ *are orthogonal at each step*. To show that p_1 is orthogonal to $Ap_0 = Ab$, simplify p_1 to cP_1: $P_1 = \|Ab\|^2 b - (b^{\mathrm{T}}Ab)\,Ab$ and $c = b^{\mathrm{T}}b/\left(b^{\mathrm{T}}Ab\right)^2$. Certainly $(Ab)^{\mathrm{T}} P_1 = 0$ because $A^{\mathrm{T}} = A$. (That simplification put α_1 into $p_1 = b - \alpha_1 Ab + \left(b^{\mathrm{T}}b - 2\alpha_1 b^{\mathrm{T}}Ab + \alpha_1^2\|Ab\|^2\right) b/b^{\mathrm{T}}b$. For a good discussion see *Numerical Linear Algebra* by Trefethen and Bau.)

Chapter 8 Linear Programming and Game Theory

Problem Set 8.1, page 381

8.1.1 The corners are at $(0, 6)$, $(2, 2)$, $(6, 0)$; see Figure 8.3.

8.1.3 The constraints give $3(2x + 5y) + 2(-3x + 8y) \leq 9 - 10$, or $31y \leq -1$. Can't have $y \geq 0$.

8.1.5 $x \geq 0$, $y \geq 0$ with added constraint $x + y \leq 0$ admits only the point $(0, 0)$.

8.1.7 The feasible set is a truncated tetrahedron bounded by the six points $(x, y, z) = (0, 0, 0)$, $(100000, 0, 0)$, $(0, 100000, 0)$, $(20000, 0, 20000)$, $(80000, 0, 20000)$, and $(20000, 60000, 20000)$. Testing $5x + 6y + 9z$ at each of these points, we find that the maximum occurs when $x = z = 20{,}000$ and $y = 60{,}000$. So 60,000 should be invested in municipals, 20,000 in federal bonds, and 20,000 in junk bonds.

8.1.9 The cost to be minimized is $1000x + 2000y + 3000z + 1500u + 3000v + 3700w$. The amounts x, y, z to Chicago and u, v, w to New England satisfy $x + u = 1{,}000{,}000$; $y + v = 1{,}000{,}000$; $z + w = 1{,}000{,}000$; $x + y + z = 800{,}000$; $u + v + w = 2{,}200{,}000$.

Problem Set 8.2, page 391

8.2.1 At present $x_4 = 4$ and $x_5 = 2$ are in the basis, and the cost is zero. The entering variable should be x_3, to reduce the cost. The leaving variable should be x_5, since $\frac{2}{1}$ is less than $\frac{4}{1}$. With x_3 and x_4 in the basis, the constraints give $x_3 = 2$, $x_4 = 2$, and the cost is now $x_1 + x_2 - x_3 = -2$.

8.2.3 The "reduced costs" are $r = \begin{bmatrix} 1 & 1 \end{bmatrix}$, so change is not good and the corner is optimal.

8.2.5 At P, $r = \begin{bmatrix} -5 & 3 \end{bmatrix}$; then at Q, $r = \begin{bmatrix} \frac{5}{3} & -\frac{1}{3} \end{bmatrix}$; R is optimal because $r \geq 0$.

8.2.7 For a maximum problem the stopping test becomes $r \leq 0$. If this fails, and the ith component is the largest, then that column of N enters the basis; the rule (Theorem 8C) for the vector leaving the basis is the same.

8.2.9 $BE = B\begin{bmatrix} \cdots & v & \cdots \end{bmatrix} = \begin{bmatrix} \cdots & u & \cdots \end{bmatrix}$, since $Bv = u$. So E is the correct matrix.

8.2.11 If $Ax = 0$ then $Px = x - A^{\mathrm{T}}\left(AA^{\mathrm{T}}\right)^{-1}Ax = x$.

Problem Set 8.3, page 399

8.3.1 Maximize $4y_1 + 11y_2$, with $y_1 \geq 0$, $y_2 \geq 0$, $2y_1 + y_2 \leq 1$, $3y_2 \leq 1$; the primal has $x_1^* = 2$, $x_2^* = 3$, the dual has $y_1^* = \frac{1}{3}$, $y_2^* = \frac{1}{3}$ with cost 5.

8.3.3 The dual maximizes yb, with $y \geq c$. Therefore $x = b$ and $y = c$ are feasible, and give the same value cb for the cost in the primal and dual; by Theorem 8F they must be optimal. If $b_1 < 0$, then the optimal x^* is changed to $(0, b_2, \ldots, b_n)$ and $y^* = (0, c_2, \ldots, c_n)$.

8.3.5 $b = \begin{bmatrix} 0 & 1 \end{bmatrix}^{\mathrm{T}}$ and $c = \begin{bmatrix} -1 & 0 \end{bmatrix}$.

8.3.7 Since $cx = 3 = yb$, x and y are optimal by Theorem 8F.

8.3.9 $x^* = \begin{bmatrix} 1 & 0 \end{bmatrix}^{\mathrm{T}}$, $y^* = \begin{bmatrix} 1 & 0 \end{bmatrix}$, with $y^*b = 1 = cx^*$. The second inequalities in both $Ax^* \geq b$ and $y^*A \leq c$ are strict, so the second components of y^* and x^* are zero.

8.3.11 (a) $x_1^* = 0$, $x_2^* = 1$, $x_3^* = 0$, $c^{\mathrm{T}}x = 3$.

(b) It is the first quadrant with the tetrahedron in the corner cut off.

(c) Maximize y_1, subject to $y_1 \geq 0$, $y_1 \leq 5$, $y_1 \leq 3$, $y_1 \leq 4$; $y_1^* = 3$.

8.3.13 As in Section 8.1, the dual maximizes $4p$ subject to $p \leq 2$, $2p \leq 3$, $p \geq 0$. The solution is $p = \$1.50$, the shadow price of protein (this is its price in steak, the optimal diet). The reduced cost of peanut butter is $\$2 - \$1.50 = 50$ cents; it is positive and peanut butter is not in the optimal diet.

8.3.15 The columns of $\begin{bmatrix} 1 & 0 & 0 & -1 & 0 & 0 \\ 0 & 1 & 0 & 0 & -1 & 0 \\ 0 & 0 & 1 & 0 & 0 & -1 \end{bmatrix}$ or $\begin{bmatrix} 1 & 0 & 0 & -1 \\ 0 & 1 & 0 & -1 \\ 0 & 0 & 1 & -1 \end{bmatrix}$.

8.3.17 Take $y = \begin{bmatrix} 1 & -1 \end{bmatrix}$; then $yA \geq 0$, $yb < 0$.

Problem Set 8.4, page 406

8.4.1 The maximal flow is 13 with the minimal cut separating node 6 from the other nodes.

8.4.3 Increasing the capacity of pipes from node 4 to node 6 or node 4 to node 5 will produce the largest increase in the maximal flow. The maximal flow increases from 8 to 9.

8.4.5 Assign capacities of 1 to all edges. The maximum number of disjoint paths from s to t then equals the maximum flow. The minimum number of edges whose removal disconnects s from t is the minimum cut. Then the maximum flow is equal to the minimum cut.

8.4.7 Rows 1, 4 and 5 violate Hall's condition; the 3 by 3 submatrix coming from rows 1, 4, 5, and columns 1, 2, 5 has $3 + 3 > 5$.

8.4.9 (a) The matrix has $2n$ 1's which cannot be covered by less than n lines because each line covers exactly two 1's. It takes n lines; there must be a complete matching.

(b) $\begin{bmatrix} 1 & 1 & 1 & 1 & 1 \\ 1 & 0 & 0 & 0 & 1 \\ 1 & 0 & 0 & 0 & 1 \\ 1 & 0 & 0 & 0 & 1 \\ 1 & 1 & 1 & 1 & 1 \end{bmatrix}$. The 1's can be covered with 4 lines; 5 marriages are not possible.

8.4.11 If each $m + 1$ marries the only acceptable man m, there is no one for #1 to marry (even though all are acceptable to #1).

8.4.13 Algorithm 1 gives 1–3, 3–2, 2–5, 2–4, 4–6 and algorithm 2 gives 2–5, 4–6, 2–4, 3–2, 1–3. These are equal length shortest spanning trees.

8.4.15 (a) Rows 1, 3, 5 only have 1's in columns 2 and 4.

(b) Columns 1, 3, 5 (in rows 2, 4)

(c) Zero submatrix from rows 1, 3, 5 and columns 1, 3, 5

(d) Rows 2, 4 and columns 2, 4 cover all 1's.

Problem Set 8.5, page 413

8.5.1 $-10x_1 + 70(1 - x_1) = 10x_1 - 10(1 - x_1)$, or $x_1 = \frac{4}{5}$, $x_2 = \frac{1}{5}$;
$-10y_1 + 10(1 - y_1) = 70y_1 - 10(1 - y_1)$, or $y_1 = \frac{1}{5}$, $y_2 = \frac{4}{5}$; average payoff $yAx = 6$.

8.5.3 If X chooses column j, Y will choose its smallest entry a_{ij} (in row i). X will not move since this is the largest entry in that row. In Problem 2, $a_{12} = 2$ was an equilibrium of this kind. If we exchange the 2 and 4 below it, no entry has this property and mixed strategies are required.

8.5.5 The best strategy for X combines the two lines to produce a horizontal line, guaranteeing this height of $\frac{7}{3}$. The combination is $\frac{2}{3}(3y + 2(1 - y)) + \frac{1}{3}(y + 3(1 - y)) = \frac{7}{3}$, so X chooses the columns with frequencies $\frac{2}{3}$, 0, $\frac{1}{3}$.

8.5.7 For columns, we want $x_1 a + (1 - x_1) b = x_1 c + (1 - x_1) d = u$ so $x_1(a - b - c + d) = d - b$. For rows, $y_1 a + (1 - y_1) c = y_1 b + (1 - y_1) d = v$ exchanges b and c. Compare u with v:

$$u = x_1(a - b) + b = \frac{(a - b)(d - b)}{a - b - c + d} + b = \frac{ad - bc}{a - b - c + d} = \text{same after } b \leftrightarrow c = v$$

8.5.9 The inner maximum is the larger of y_1 and y_2; x concentrates on that one. Subject to $y_1 + y_2 = 1$, the minimum of the *larger* y is $\frac{1}{2}$. Notice $A = I$.

8.5.11 $Ax^* = \begin{bmatrix} \frac{1}{2} & \frac{1}{2} \end{bmatrix}^{\mathrm{T}}$ and $yAx^* = \frac{1}{2}y_1 + \frac{1}{2}y_2 = \frac{1}{2}$ for all strategies of Y; $y^*A = \begin{bmatrix} \frac{1}{2} & \frac{1}{2} & -1 & -1 \end{bmatrix}$ and $y^*Ax = \frac{1}{2}x_1 + \frac{1}{2}x_2 - x_3 - x_4$, which cannot exceed $\frac{1}{2}$; in between is $y^*Ax^* = \frac{1}{2}$.

8.5.13 Value 0 (fair game). X chooses 2 or 3, y chooses odd or even: $x^* = y^* = \left(\frac{1}{2}, \frac{1}{2}\right)$.

Appendixes

Problem Set A, page 420

A.1 (a) The largest $\dim(\mathbf{S} \cap \mathbf{T})$ is 7, when $\mathbf{S} \subset \mathbf{T}$.

(b) The smallest $\dim(\mathbf{S} \cap \mathbf{T})$ is 2.

(c) The smallest $\dim(\mathbf{S} + \mathbf{T})$ is 8, when $\mathbf{S} \subset \mathbf{T}$.

(d) The largest $\dim(\mathbf{S} + \mathbf{T})$ is 13 (all of $\mathbf{R}^{13}$).

A.3 $\mathbf{V} + \mathbf{W}$ and $\mathbf{V} \cap \mathbf{W}$ contain the matrices $\begin{bmatrix} a_{11} & a_{12} & a_{13} & a_{14} \\ a_{21} & a_{22} & a_{23} & a_{24} \\ 0 & a_{32} & a_{33} & a_{34} \\ 0 & 0 & a_{43} & a_{44} \end{bmatrix}$ and $\begin{bmatrix} a_{11} & a_{12} & 0 & 0 \\ 0 & a_{22} & a_{23} & 0 \\ 0 & 0 & a_{23} & a_{34} \\ 0 & 0 & 0 & a_{44} \end{bmatrix}$.

$\dim(\mathbf{V} + \mathbf{W}) = 13$ and $\dim(\mathbf{V} \cap \mathbf{W}) = 7$; add to get $20 = \dim \mathbf{V} + \dim \mathbf{W}$.

A.5 The lines through $(1, 1, 1)$ and $(1, 1, 2)$ have $\mathbf{V} \cap \mathbf{W} = \{0\}$.

A.7 One basis for $\mathbf{V} + \mathbf{W}$ is v_1, v_2, w_1; $\dim(\mathbf{V} \cap \mathbf{W}) = 1$ with basis $(0, 1, -1, 0)$.

A.9 The intersection of column spaces is the line through $y = (6, 3, 6)$:

$$y = \begin{bmatrix} 1 & 5 \\ 3 & 0 \\ 2 & 4 \end{bmatrix} \begin{bmatrix} 1 \\ 1 \end{bmatrix} = \begin{bmatrix} 3 & 0 \\ 0 & 1 \\ 0 & 2 \end{bmatrix} \begin{bmatrix} 2 \\ 3 \end{bmatrix} \text{ matches } \begin{bmatrix} A & B \end{bmatrix} x = \begin{bmatrix} 1 & 5 & 3 & 0 \\ 3 & 0 & 0 & 1 \\ 2 & 4 & 0 & 2 \end{bmatrix} \begin{bmatrix} 1 \\ 1 \\ -2 \\ -3 \end{bmatrix} = 0$$

The column spaces have dimension 2. Their sum and intersection give $3 + 1 = 2 + 2$.

A.11 $F_2 \otimes F_2 = \begin{bmatrix} F_2 & F_2 \\ F_2 & -F_2 \end{bmatrix} = \begin{bmatrix} 1 & 1 & 1 & 1 \\ 1 & -1 & 1 & -1 \\ 1 & 1 & -1 & -1 \\ 1 & -1 & -1 & 1 \end{bmatrix}$.

A.13 $A_{3D} = (A_{1D} \otimes I \otimes I) + (I \otimes A_{1D} \otimes I) + (I \otimes I \otimes A_{1D})$.

Problem Set B, page 427

B.1 $J = \begin{bmatrix} 2 & 0 \\ 0 & 0 \end{bmatrix}$ (A is diagonalizable); $J = \begin{bmatrix} 0 & 1 & 0 \\ 0 & 0 & 0 \\ 0 & 0 & 0 \end{bmatrix}$ (eigenvectors $(1, 0, 0)$ and $(2, -1, 0)$)

B.3 $e^{Bt} = \begin{bmatrix} 1 & t & 2t \\ 0 & 1 & 0 \\ 0 & 0 & 0 \end{bmatrix} = I + Bt$ since $B^2 = 0$. Also $e^{Jt} = I + Jt$.

B.5 $J = \begin{bmatrix} 1 & 0 & 0 \\ 0 & 4 & 0 \\ 0 & 0 & 6 \end{bmatrix}$ (distinct eigenvalues); $J = \begin{bmatrix} 0 & 1 \\ 0 & 0 \end{bmatrix}$ (B has $\lambda = 0, 0$ but rank 1).

CPSIA information can be obtained
at www.ICGtesting.com
Printed in the USA
FFOW01n2154290616
25521FF

9 780495 013259